PHYSICO-CHEMICAL CHARACTERISATION
OF PLANT RESIDUES
FOR INDUSTRIAL AND FEED USE

Proceedings of a workshop held in Aberdeen, Scotland (UK) from 21 to 23 June 1988 under the auspices of COST (European Cooperation in Scientific and Technological Research)—COST 84-bis, organised with the support of the Commission of the European Communities by the Rowett Research Institute, Aberdeen.

PHYSICO-CHEMICAL CHARACTERISATION OF PLANT RESIDUES FOR INDUSTRIAL AND FEED USE

Edited by

A. CHESSON and E. R. ØRSKOV

Rowett Research Institute, Aberdeen, Scotland, UK

ELSEVIER APPLIED SCIENCE
LONDON and NEW YORK

ELSEVIER SCIENCE PUBLISHERS LTD
Crown House, Linton Road, Barking, Essex IG11 8JU, England

Sole Distributor in the USA and Canada
ELSEVIER SCIENCE PUBLISHING CO., INC.
655 Avenue of the Americas, New York, NY 10010, USA

WITH 29 TABLES AND 73 ILLUSTRATIONS

© 1989 ECSC, EEC, EAEC, BRUSSELS AND LUXEMBOURG

British Library Cataloguing in Publication Data
Physico-chemical characterisation of plant residues for
industrial and feed use.
1. Plants. Cells. Walls. Biochemistry
I. Chesson, A.   II. Ørskov, E. R. (Egil Roberts),
1934–
581.8'75

ISBN 1-85166-376-2

Library of Congress CIP data applied for

Publication arrangements by Commission of the European Communities, Directorate-General
Telecommunications, Information Industries and Innovation, Scientific and Technical Communication Service, Luxembourg

EUR 11942

LEGAL NOTICE

Special regulations for readers in the USA

Printed in Northern Ireland at The Universities Press (Belfast) Ltd.

CONTENTS

INTRODUCTION

The workshop reported in this volume is one of a series sponsored by the Commission of the European Communities, Directorate-General for Science, Research and Development (DG XII), under the Concerted Action Programme COST 84-bis, entitled "Use of lignocellulose containing by-products and other plant residues for animal feeding". Since COST 84-bis was established there has been a major shift of emphasis in agricultural research in Europe, with the development of alternative uses for crops and their by-products becoming a priority issue. In recognition of this recent workshops held under the aegis of COST 84-bis have been equally concerned with the potential of lignocellulosic residues to form the feedstock for a variety of new and established industrial uses in addition to their established use as animal feed.

Development of strategies for the use of plants or plant residues with a high cell wall content is dependant on knowledge of cell wall structure and organisation and how structure relates to the behaviour of the wall during mechanical, chemical or biological processing.

Progress in cell wall research has been greatly facilitated by the substantial developments in methods of instrumental analysis that have occurred during the last decade. Plant tissues now can be examined in far greater detail and far more rapidly than was hitherto possible, often without the need for extraction or modification of the cell wall or its component polymers. The use of instrumental techniques in biological investigations is not, of course, a recent innovation. Research on the structure and properties of plant cell walls has long made use of light and electron microscopy to describe and visualise plant tissues and their individual cells. However such investigation have been largely qualitative in nature. The addition of an analytical capacity to both light and electron microscopes (e.g. energy dispersive X-ray analysis) coupled with developments in cytohistochemical techniques and image analysis have made routine the use of microscopic methods in quantitative investigations. Newer developments, such as tunnelling microscopy, are likely to add a further dimension to the resolution offered by microscopy enabling the organisation of plant cell wall surfaces to be examined at a molecular level.

Chemical techniques used to establish the composition and organisation of cell walls also have been considerably aided by developments in separation methods (high performance liquid chromatography, capillary gas chromatography and supercritical fluid chromatography) and their associated detection methods, particularly mass-spectrometry. However all such methods involve, at some stage, the disruption and fragmentation of the wall with a consequent loss of structural information. Developments in spectroscopic analysis, notably those based on solid-state nuclear magnetic resonance spectroscopy, which allow the sample to be examined in its native state are beginning to supersede many of the established chemical methods. Improvements in the resolution offered by NMR imaging techniques also have reached a stage when the "NMR microscope" seems a real possibility. With this technique dynamic processes, such as cell wall development or degradation could be followed over long periods in intact tissues.

Even when applied to homogeneous cell wall samples (i.e. derived from a single cell type), chemical techniques provide only an average picture of composition and organisation. Biological evidence suggests that this is an over-simplification and points to a considerable variation in both composition and architecture within a single wall. The resolution needed to detect such variation can only be provided directly or indirectly by instrumental techniques. Depth profiling using methods such as photoacoustic spectroscopy offers scope for future research in this area. Similarly instrumental methods developed to examine the chemistry and properties of surfaces (X-ray photoelectric spectroscopy, laser ablation-time of flight mass spectrometry) are beginning to be applied to biological systems including studies of the important changes which occur at cell wall surfaces during their degradation.

The papers included in this volume do not, by any means, cover all of the instrumental techniques potentially available to research workers concerned with plant cell walls and their uses. This would not be feasible within a single volume and at a time of such rapid development. The contributions do, however, serve to introduce many of the major techniques and show how such methods may contribute to a better understanding of the use of plant cell walls as a resource, not only for animal feed, but as a raw material for a wide range of industrial processes.

USE OF MODERN NMR SPECTROSCOPY IN PLANT CELL WALL RESEARCH

D.S. HIMMELSBACH

R.B. Russell Agricultural Research Center, Agricultural Research Service, United States Department of Agriculture, P.O. Box 5677, Athens, Georgia 30613, U.S.A.

SUMMARY

Three areas of nuclear magnetic resonance (NMR) spectroscopy are discussed that are having a significant impact in plant cell wall research. These are: (1) high field/high resolution solution methods, especially 2D methods, on isolated components; (2) solid-state techniques, especially using cross-polarization/ magic-angle spinning (CPMAS) and (3) magnetic resonance imaging (MRI), both macro and micro. Examples are taken from recent literature and from our own laboratory. Some additional data interpretations are made. A few comparisons are made with other methods to explain features or relate results.

INTRODUCTION

The increased commercial availability of high field, 11.5T and 14T (500 and 600 MHz for protons), NMR superconducting magnets has increased signal dispersion and sensitivity that is obtainable in high resolution solution studies. Improved two-dimensional (2D) pulse sequences have permitted the unraveling of complex polymeric structures, like those so often encountered in plant cell walls. These techniques permitted definition of conformation and configuration. Correct spectral assignments can now be made directly by instrumental techniques rather than by comparison to reference spectra. The perfection of inverse detection techniques has permitted the extension of 2D techniques with low-γ-nuclei (such as carbon-13 or nitrogen-15) to hard to isolate or sparing soluble macromolecules.

Solid-state NMR techniques, which allow the study of the entire intact plant cell wall, have been perfected. The most notable method combines the techniques of cross-polarization and magic-angle spinning (CPMAS) for the observation of carbon signals on the relaxation time scale of the protons. The routine attainment of higher spinning rates (up to 17 KHz) has permitted the observation of line widths on the order of solution spectra and the use of higher field magnets for obtaining carbon spectra. High power multipulse techniques permit the observation of solid-state proton spectra with narrower line widths, giving higher resolution.

Magnetic resonance imaging (MRI) is a new technique as far as studies on cell walls of plants are concerned. Development of this technique has been mostly spurred by the medical profession and, as a result, the emphasis has been on human or animal tissue and not plant tissue. The most recent change in this area is from macro (whole body) to micro (cell level) techniques. Micro-imaging offers the greatest opportunities for cell wall research, particularly the possibilities of studying totally intact viable tissue or, dynamically, the degradation of cell walls.

SOLUTION NMR

The use of solution NMR in the study of plant cell walls has involved the study of extracted components of the walls rather than the entire wall itself, due to its general insolubility. These studies have generally involved the study of lignin and polysaccharide components. Figure 1 shows the 125 MHz carbon spectrum of wheat straw lignin as extracted by 96% dioxane and treated with a cellulase to further remove carbohydrate[1]. These spectra show considerably increased signal dispersion with the use of the higher field instrument. The confirmation of the number of attached protons to each carbon was obtained

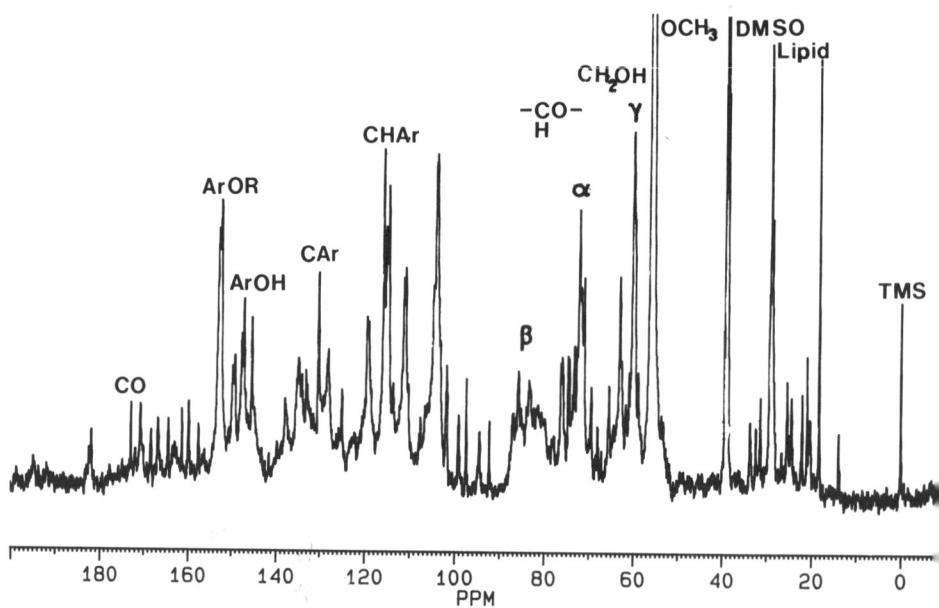

Figure 1. One-dimensional 125 MHz ^{13}C high-resolution solution spectra of wheat straw lignin extracted by 96% (v/v) dioxane-water.

by the use of the 1D multiplicity sorting technique of Distortionless Enhancement by Polarization Transfer (DEPT)[2]. By employing isotopic ^{13}C enrichment, carbon-carbon connectivities have been obtainable on lignin using the 2D Incredible Natural Abundance Double Quantum Experiment (INADEQUATE)[3]. This experiment, although information packed, is extremely time consuming when one must rely on detection at natural isotopic abundance of the carbon nucleus. This kind of connectivity experiment will probably be subplanted by the inverse techniques like HMQC (Heteronuclear Multiquantum Connectivity)[4] and HMBC (Heteronuclear Multiple-bond Connectivity[5]. Examples of the use of these techniques are shown in Figures 2 and 3 respectively for an important sucrose ester isolated from tobacco leaves[6]. These techniques have also been successfully applied to plant cell wall polysaccharides such as xyloglucans[7]. The HMQC experiment's utility is in making assignments between the two 1D domains. The HMBC experiment permits sequencing of residues by being able to hop over other heteroatoms, such as oxygen, with the long-range coupling. Once an unambiguous starting point is established the whole molecule can often be traced from bond to bond. As these techniques become more common place, we should have much re-evaluation of older solution spectra of cell wall materials such as lignin and hemicellulose.

SOLID-STATE NMR

The cross-polarization/magic angle spinning (CPMAS) technique with carbon observation has been applied to both isolated fractions of the cell wall (cellulose (Figure 4a), hemicellulose (Figure 4b), lignin-carbohydrate complex (Figure 5a) and lignin (Figure 5b)) plus the cell walls itself, as neutral detergent fibre residue (Figure 6)[8]. It has been possible, with solid-state ^{13}C NMR to follow the ruminant digestion of cell wall material. The spectra of the neutral detergent residues resulting from *in sacco* digestion of a mixed silage are shown in Figure 7[9]. As digestion proceeds, carbohydrate and protein decrease while phenolic and lignin materials increase their contribution to the residue.

These NMR investigations of silage digestion have been supported by mid-infrared (MIR, Figure 8) and near-infrared (NIR, Figure 9), evaluation. All of these spectroscopic techniques show similar results that are in agreement with the wet chemical gravimetric data (Table 1).

Some components are more easily discernible by one technique than another. Lignin or polyphenols are most easily discernable in the NMR spectra in the 140-160 ppm region, indicative of phenolic hydroxyl, and at 56 ppm, indicative of the aromatic methoxyl group. They are much harder to pick out of the MIR spectra, but are detectable by the aromatic C=C stretch at 1505 cm^{-1}, C-O stretches at 1235 and 1265 cm-1 and by the O-H deformation at 1034 cm-1.

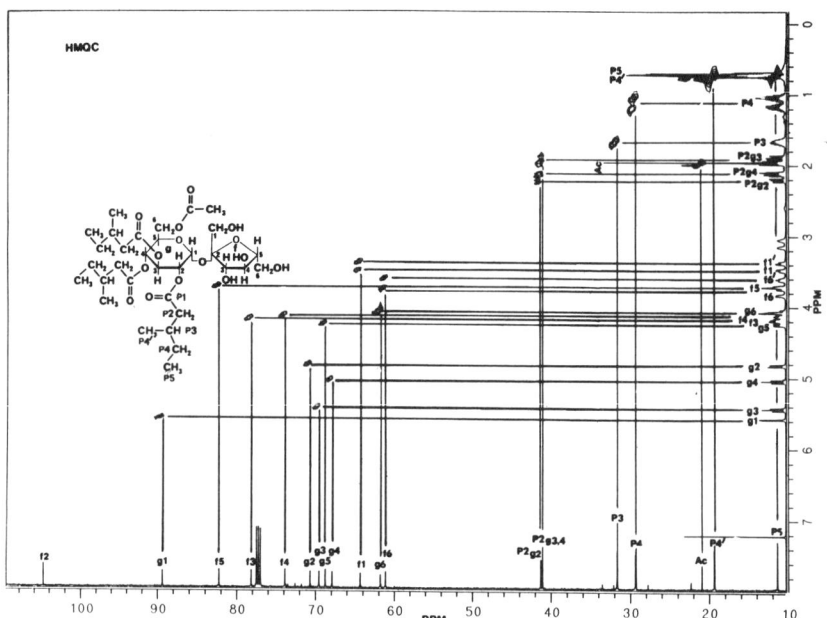

Figure 2. Inverse (¹H) detected HMQC (one-bond C-H connectivity) plot of 50 mgs of a sucrose ester in CDCl₃ at 500 MHz for proton and 125 MHz for carbon. The one-bond coupling ($^1J_{CH}$) was set at 125 Hz.

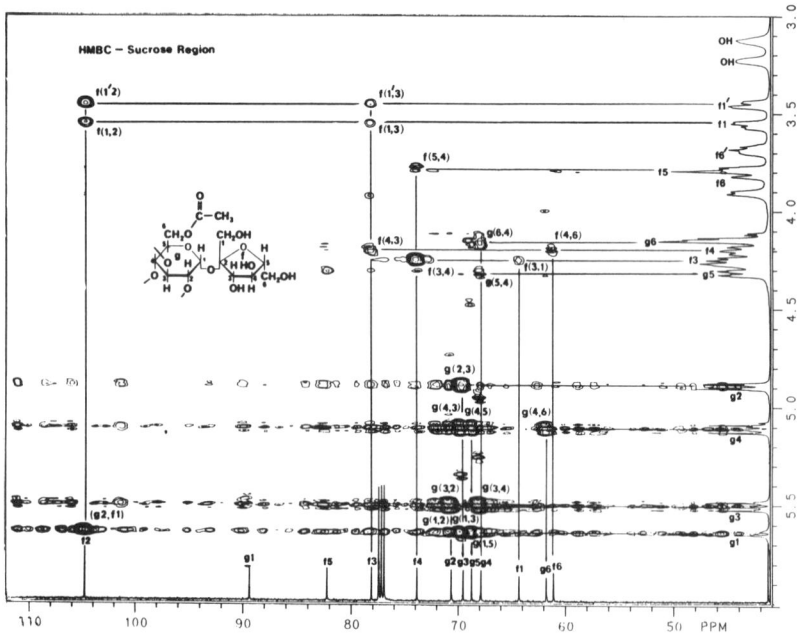

Figure 3. Inverse (¹H) detected HMQC (two- and three-bond C-H connectivity) plot of 50 mgs of a sucrose ester at 500 MHz for protons and 125 MHz for carbons. The multiple-bond coupling constant was set to 7 Hz.

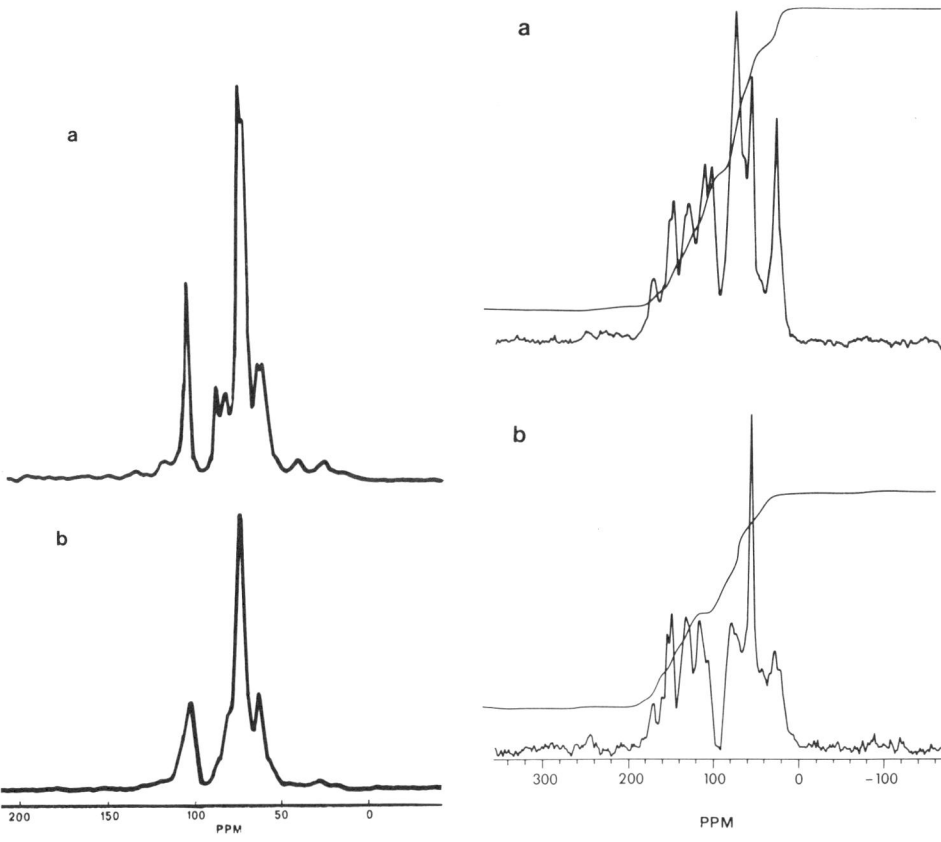

Figure 4. CPMAS ^{13}C NMR spectra of: a) cellulose, b) hemicellulose

Figure 5. CPMAS ^{13}C NMR of: a) lignin-carbohydrate complex from cell walls, as neutral detergent fibre (NDF), b) lignin extracted from whole grass from 'coastal' Bermudagrass.

They can just be discerned at 2264 nm in the NIR spectrum, due to the combination band between the CH and C=C stretch frequencies and at 2134 nm due to the O-H and C-O stretch combination band. In all spectroscopic techniques and in the wet chemical data, phenolics increase in the residue with time of digestion. The split signal in the NMR spectrum gives some further information; the signal at 153 ppm indicates that syringyl units are preferentially retained. This is consistent with the rapid increase in methoxyl signal. Lipids and waxes, apparently from cuticle, also increase in the cell wall residues. They are easily discernable at 29 and 22 ppm in the NMR spectra, at 2930 and 2850 cm^{-1} plus 1580 and 1530 cm^{-1} in the MIR and at 1724 and 1758 nm plus 2302 and 2344 nm in the NIR.

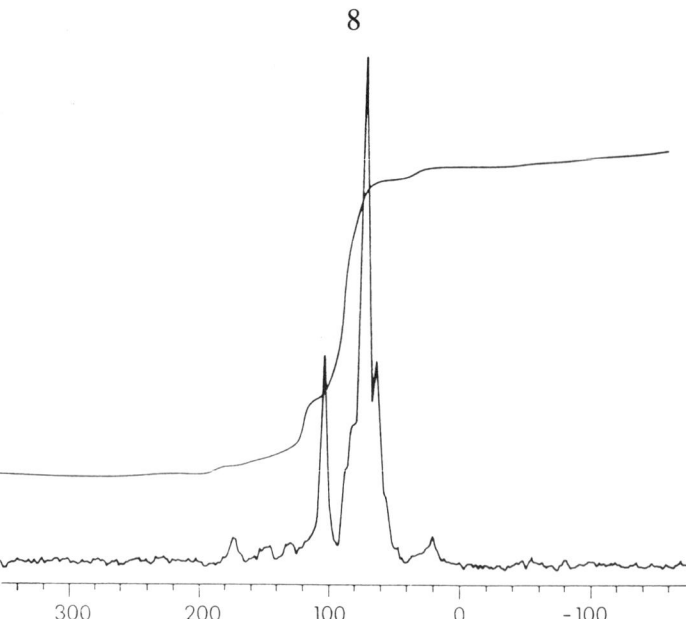

Figure 6. CPMAS ^{13}C NMR spectrum of cell walls (as neutral detergent fibre residue) from 'coastal' Bermudagrass.

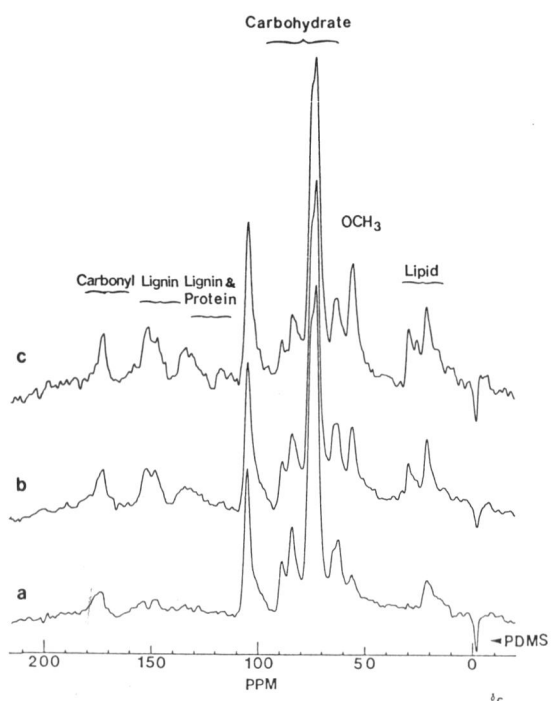

Figure 7. CPMAS ^{13}C NMR spectra of neutral detergent fibre residues from 21% crude fibre grass silage. The spectra are residues taken at: a) 0 hr, b) 48 hr, and c) 336 hr of digestion, *in sacco*.

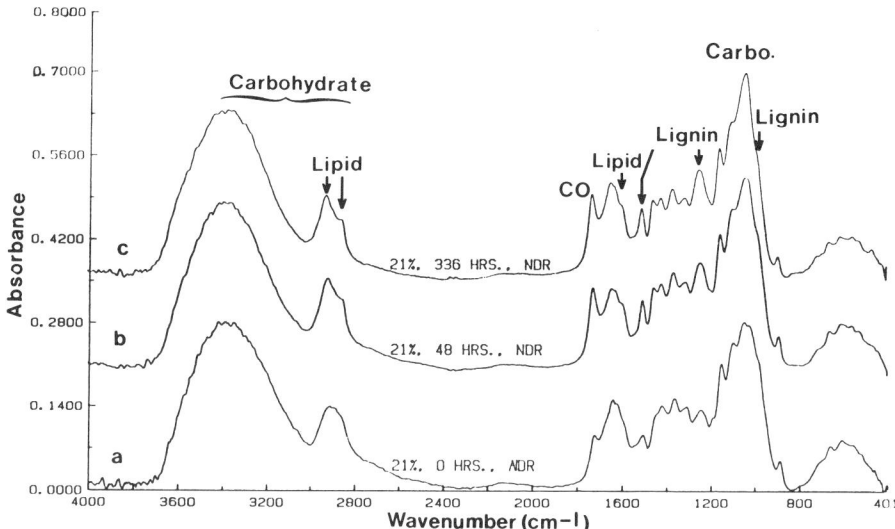

Figure 8. Fourier transforms mid-infrared (MIR) spectra of neutral detergent residues from a 21% crude fibre silage at: a) 0 hr, b) 48 hr and c) 336 hr of digestion.

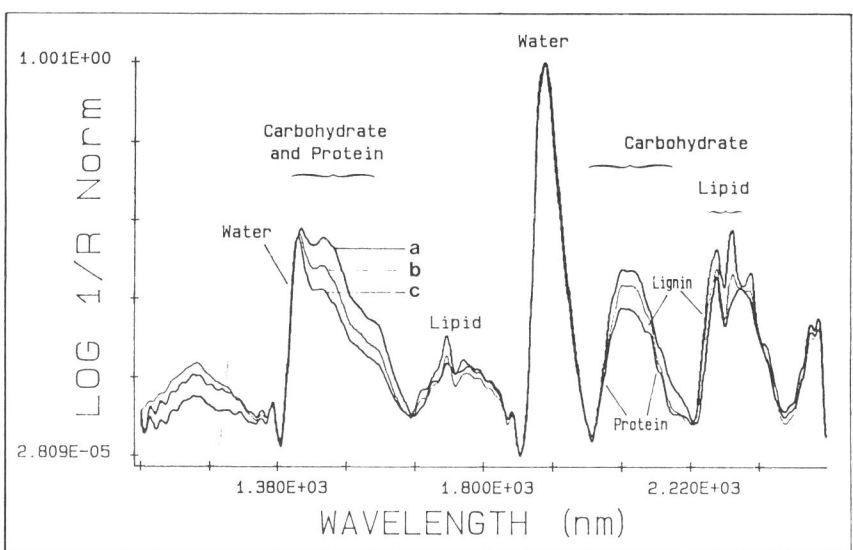

Figure 9. Near-infrared (NIR) diffuse reflectance spectra of neutral detergent residues from a 21% crude fibre silage at: a) 0 hr, b) 48 hr and c) 336 hr of digestion.

TABLE 1. Wet chemical data for 21% crude fibre silage

Sample	CP	ADiP	NDF	ADF	Lignin
UNTR	20.59	0.675	41.17	24.59	1.65
0 hr	15.52	0.631	69.81	40.36	3.05
12 hr	13.13	0.881	71.64	42.20	4.52
24 hr	10.66	1.119	70.23	41.11	7.60
48 hr	10.19	1.425	65.77	37.71	10.20
336 hr	12.26	2.138	51.57	31.05	13.28

CP, crude protein; AdiP, acid detergent insoluble protein; NDF, neutral detergent fibre; ADF, acid detergent fibre

Carbohydrates decrease on a relative basis to the other components in the NMR spectra as observed in the region of 105-62 ppm. The signals at 89 and 64 ppm, when compared to those at 84 and 62 ppm, give an indication of relative crystallinity of the cellulosic fraction. The decrease in the signal at 84 ppm which indicates perferential digestion of amorphous carbohydrates may be hemicellulose or amorphous cellulose. In the MIR spectra the loss of carbohydrate can be noted in the O-H and C-H stretching area (about 3400-2500 cm^{-1}) plus in the O-H deformation region (1200-800 cm^{-1}). In the NIR spectra the loss of carbohydrate is very discernable in the regions from 1480-1580 nm and from 2080-2120 nm. Protein is also a component that generally decreases with digestion. The protein signals are not distinguishable from lignin in this cell wall material, either in the NMR or NIR spectra. They overlap carbohydrate in the NIR spectra; however, two protein bands are barely discernable at 2054 and 2168 nm.

Overall, in the spectral arena, NMR seems to give the most definitive *in situ* information, especially in following phenolic constituents that are difficult to follow by any method. However, NIR effects some possibilities. Lipid or waxes can be detected by all three methods, but NMR and NIR appear to be the most sensitive to their presence. No method is optimal for all components only a combination of techniques yield adequate total information.

MAGNETIC RESONANCE IMAGES

Magnetic resonance imaging has been developing in the last 10 years. Macro (i.e. whole body) type imaging has been extremely useful in the medical profession. This scale of imaging is not very applicable to the type of detailed information that is required to study plant cell walls. The achievement of submillimeter spatial resolution is required for this type of work. This has been termed micro-imaging[10]. Instrumentally it requires smaller samples, high magnetic fields and rapidly changeable linear gradients. At present the best spatial resolution obtainable with proton images has been 10 μm[11]. Plates 1 and 2 show applications to lignocellulosic materials. Plate 1 shows the cross section of a tree stem that details the structure of the sapwood, heart and earlier branch[12]. Plate 2 shows the image of a leaf blade with the detail of light microscopy[12]. Most of the NMR imaging has concentrated on proton imaging of water or oil. NMR imaging methods are now being extended to other abundant nuclei and solids. One can obviously see that this technique leads to numerous applications for cell wall structure determination and only awaits availability to the researcher.

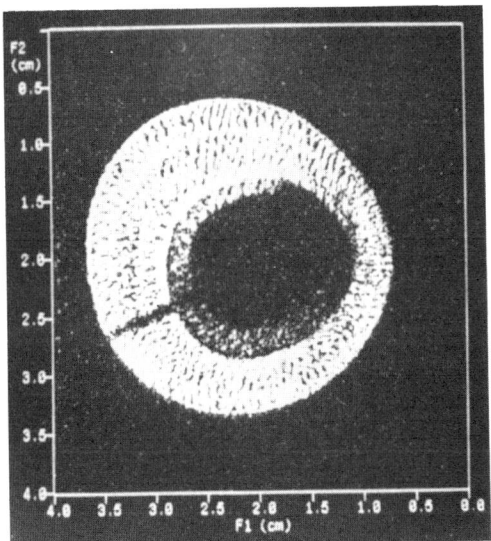

Plate 1. Cross section of tree stem taken using multislice acquisition.

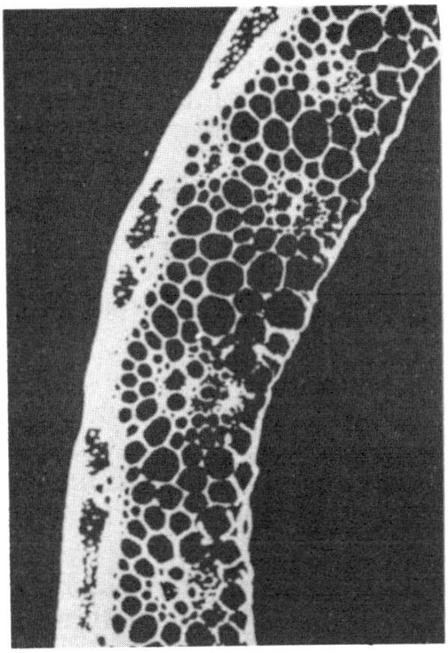

Plate 2. Proton image cross section of leaf blade.

CONCLUSION

NMR is an extremely valuable and sensitive tool for the study of plant cell wall structure and composition. The future use of this technique seems almost boundless, limited only by our ability to conceive suitable experiments. When information from NMR is combined with other techniques, even greater synergistic advances can be made in the area of cell wall research.

REFERENCES

1. Jung, H.J.G. and Himmelsbach, D.S. (1988). Isolation and characterization of wheat straw-lignin. *Journal of Agricultural and Food Chemistry* (In Press).

2. Doddrell, D.M.; Pegg, D.T. and Bendall, M.R. (1982). Distortionless enhancement of NMR signals by polarization transfer. *Journal of Magnetic Resonance* **48**: 323-327.

3. Lapierre, C.; Monties, B.; Guittet, E. and Lallemeard, J.T. (1987). Two-dimensional carbon-13 NMR of popular lignins; Study of carbon connectivities and examination of signal assignments by means of the INADEQUATE technique. *Holzforschung* **41** (1): 51-58.

4. Bax, A. and Subramanian (1986). Sensitivity-enhanced two-dimensional heteronuclear shift correlation NMR spectroscopy. *Journal of Magnetic Resonance* **67**: 565-569.

5. Bax, A. and Summers (1986). ^1H and ^{13}C assignments from sensitivity-enhanced detection of hetero-nuclear multi-bond connectivity by 2D multiple quantum NMR. *Journal of the American Chemical Society* **108**: 2093-2094.

6. Himmelsbach, D.S.; van Halbeek, H. and Arrendale, R.F. (1988). Assignment of proton and carbon-13 spectra of sugar esters by inverse detected 2D NMR spectroscopy. Abstracts 3rd Chemical Congress of North America, Toronto, June 1988, **1:** 38.

7. van Halbeek, H.; York, W.S.; Darvill, A.G. and Albersheim, F. (1988). Application of ^1H-detected (^1H, ^{13}) shift-correlation and (^1H, ^1H) Hartment-Hahn spectroscopy to the study of plant cell wall xylogolucan oligosaccharides, Abstracts 3rd Chemical Congress of North America, Toronto, June 1988, **1:** 74.

8) Barton, F.E. II; Akin, D.E.; Windham, W.R. and Himmelsbach, D.S. (1983). Methods of forage analysis-quantitative and qualitative aspects. In: *Wood and Agricultural Residues,* Soltes, E.J. ed. Academic Press, New York. p. 167-202.

9) Himmelsbach, D.S.; Boer, H.; Akin, D.E. and Barton, F.E. II. Solid-state carbon-13 NMR, FTIR, and NIRS spectroscopic studies of ruminant silage digestion. Proceedings of the Royal Society of Chemistry: Spectroscopy Across the Spectrum, Norwich, July 1987.

10. Kuhn, W. (1987). The NMR-Microscope. Applications and Limitations. Bruker Report 1/1987, 40-41.

11. Aguayo, J.B.; Blackband, S.J.; Schoenvger, J.; Mattingly, M.A. and Hintermaud, M. (1986). Nuclear magnetic resonance imaging of a single cell. *Nature* **322**: 190-191.

12. Courtesy of David L. Foxall, Spectroscopy Imaging System Co., Fremont, CA, U.S.A.

NEAR- AND MID-INFRARED STUDIES OF THE CELL WALL STRUCTURE OF CEREAL STRAW IN RELATION TO ITS RUMEN DEGRADABILITY

J.D. RUSSELL[1], I. MURRAY[2] and A.R. FRASER[1]

[1]*Macaulay Land Use Research Institute, Craigiebuckler, Aberdeen AB9 2QJ, U.K.*
[2]*North of Scotland College of Agriculture, 581 King Street, Aberdeen AB9 1UD, U.K.*

SUMMARY

Mid-infrared reflection studies using a multiple internal reflection technique have unambiguously distinguished inner and outer surfaces of cereal and rape straws, indicating significantly different chemical compositions. These chemical differences are important in controlling the progress of straw degradation in sheep rumen. The reflection technique analyses a surface layer up to 10 μm thick and has shown that for cereal straw the outer layers containing waxy cuticle and amorphous silica undergo little or no degradation over prolonged incubation. The inner layers, carbohydrate-rich and containing very little lignin, undergo extensive degradation of carbohydrate and accumulation of lignin. This accumulation terminated degradation after about 48 h. In contrast, the outer layers of rape straw contain a deposit of calcite and very little cuticular wax. The acidic conditions in the rumen dissolved the calcite, and carbohydrate in both surfaces degrades simultaneously until, as for cereal straw, lignin accumulation intervenes.

The near-infrared diffuse reflectance (NIR) technique has been used to assess degradability of cereal straw directly. NIR spectra of twenty-five cereal straws and cereal straw fractions were correlated with their dry matter degradability determined by 48 hour incubation in nylon bags in the rumen of sheep. One wavelength segment could account for 89% of the variance in 48 hour degradability giving a standard error of estimate of 5% units.

NIR spectra of leaf and stem fractions of cereal straws were appreciably different over the regions 1600-1700 and 2200-2300 nm. These same regions were analytically useful in estimating degradability and are well established for testing silage for "*in vivo*" DOMD. Chemically extracted straw lignins show absorptions in these regions which are negatively correlated to degradability.

A similar relationship between degradability and absorbance of the C=C lignin band at about 1500 cm[-1] was observed in the mid-IR range, supporting the inverse relationship between lignin content and degradability of cereal straw.

INTRODUCTION

Conventional chemical analysis has proved inadequate for evaluating the digestibility of cereal straws. In rumen degradation studies, wet chemical analysis of whole plant particles can provide only limited information on the chemistry of the surface layers of the particles. It is at these surfaces however that degradation by colonising rumen microorganisms occurs, and it is therefore important to try to address this area by direct physico-chemical means in order that changes in chemistry and composition may be monitored and understood. One such technique, multiple internal reflectance or attenuated total reflectance (ATR) spectroscopy in the mid-infrared (MIR) range can provide spectral information for surface layers up to about 5-10 μm thick, and requires little or no sample preparation.

The information provided by this method in straw degradation work is necessarily qualitative and of a comparative nature. When quantitative determination of chemical components in cereal straws is required, conventional chemical analysis has proved inadequate, for example in evaluating the digestibility of ceral straws. Near-infrared (NIR) reflectance spectroscopy offers the possibility of providing this information rapidly and cost-effectively. NIR reflectance spectra of plant materials are smooth and rather featureless, comprising overlapping absorption arising mainly from overtone and combination bands of MIR fundamentals. Absorption due to CH, OH and NH is most readily recognised, but at present only tentative assignments can be made by analogy with spectra of pure compounds. The shape of the spectra contain compositional information which can be abstracted by statistical calibration using correlation transform spectroscopy.

This paper describes the application of these MIR- and NIR-reflectance techniques to the mechanism of rumen digestion of cereal straw and the direct measurement of its degradability.

MATERIALS AND METHODS

Preparation of samples

Straws were chopped into short 1-2 cm lengths and the fractions passing through a 4 mm sieve but retained by a 2 mm sieve were used in the spectroscopic studies. For MIR reflectance work the straws received no further treatment. For NIR reflectance, they were ball milled.

Rumen digested straw (1-2 cm x 2-4 mm) was recovered from nylon bags of 5 μm pore size, after incubation for up to 120 h in the rumen of sheep fed on grass *ad libitum*.[1] One such nylon bag from each straw was washed with only enough water to remove coarse, adhering rumen contents before freeze drying.

The straws examined by MIR were of wheat (Timor) and Barley (Golden Promise). For the NIR study a set of 25 straws and straw fractions was examined (Table 1). For these samples, the 48 h degradability was chosen for comparison with the spectroscopic data.

TABLE 1. Cereal straw samples and fractions

No.	Type	Variety	Fraction	Degradability (48 h)
1	SB	Corgi	Leaf Blade	81.59
2	SB	Corgi	Whole	58.25
3	SB	Corgi	Stem	43.29
4	SB	Corgi	Leaf Sheath	75.01
5	SB	Corgi	Chaff	57.88
6	SB	Celt	Whole	45.57
7	SB	Doublet	Whole	57.86
8	SB	G. Promise	Whole	41.49
9	SB	Golf	Whole	42.60
10	SB	Heriot	Whole	50.67
11	SB	Klaxon	Whole	39.97
12	WW	Aquila	Leaf	63.95
13	WW	Aquila	Stem	30.74
14	WW	Brigand	Leaf	58.23
15	WW	Brigand	Stem	29.62
16	WW	Longbow	Leaf	59.85
17	WW	Longbow	Stem	31.01
18	Oat	Cabana	Leaf	53.10
19	Oat	Trafalgar	Leaf	48.40
20	W	Aquila	Leaf	63.90
21	W	Brigand	Leaf	58.20
22	Oat	Ballad	Internode	25.40
23	Oat	Dula	Internode	27.10
24	W	Boxer	Internode	30.40
25	W	Renard	Internode	31.50

Spectroscopy

MIR. Reflectance spectra were obtained in the range 4000-300 cm^{-1} on a Perkin Elmer 580B spectrometer using a SPECAC accessory with a 25-reflection KRS-5 crystal, set at an optimum angle of 60° for the incident radiation. Strips of straw approximately 10 mm x 5 mm were arranged with appropriate surface (inner or outer) in contact with both faces of the crystal. Coverage of about 80% and moderate pressure were required to give reasonable spectral intensity. The spectrum of the blank crystal was recorded after removal of each sample and used as a background blank for the next sample. An interfaced microcomputer and in-house programmes were used to effect background subtraction, base-line flattening and other essential spectral manipulation[2].

For normal transmission spectra in the MIR, straw samples were pre-ground after moistening with a few drops of isopropyl alcohol, then 0.8 mg was incorporated with 170 mg KBr in a 12.5 mm diameter pressed disk.

NIR.   Reflectance spectra were obtained by scanning the samples as a dry-packed powder 1 cm deep in a quartz window cuvette.   Spectra were recorded on a Pacific Scientific Co. Model 6100 coupled to a Digital Corporation PDP 11/03 computer using the software of Shenk *et al*[3].   Reflectance spectra were stored as log ^{1}/R at 2 nm intervals from 1100-2500 nm.   Stepwise multiple linear regression (SMLR) was used to select wavelength segments which explain the degradability data in an equation of the form:

Predicted 48 h degradability % = $B_o + B_1W_1 + B_2W_2 + B_3W_3$ ....

where B values are constants and W values are optical data at wavelengths λ_1, λ_2, ... Graphs of the correlation coefficient versus wavelength and the standard deviation of spectra were used to assess analytically useful wavelengths.

RESULTS AND DISCUSSION

MIR.   The ATR spectra of wheat straw have been selected from the work of Russell *et al*.[4] to illustrate the type of information this reflectance technique can provide.   Distinct differences are detected between spectra of inner and outer surfaces of untreated straw (Figure 1a), indicating that these surface layers have substantially different chemical compositions.   The spectrum of the outer layer exhibits the aliphatic C-H stretching bands at 2850 and 2920 cm^{-1} and stretching of C=O ester at 1725 cm^{-1} of cutin from the waxy cuticular layer, and strong Si-0 bands at 465, 797 and 1505-1200 cm^{-1} of amorphous silica (SiO_2).   The major component common to both layers is polysaccharide with cellulose and hemicellulose characteristics being detectable.   Lignin absorption bands at 1050, 1455, 1420, 1233 and 835 cm^{-1} are very weak in the inner surface spectrum, but are scarcely detectable in the outer, possibly because the lignin lies below the level at which SiO_2 occurs and may not be penetrated by the radiation.   Very weak acetyl absorption at 1725 cm^{-1} was detectable in the inner layer.   It is probably present in the outer also but the band is overlapped by cutin carbonyl at the same frequency.   The assignment of the weak 1725 cm^{-1} band to acetyl in the inner surface layer spectrum (Figure 1 a,i) was confirmed by its absence in the spectrum of alkali treated straw inner surface (Figure 1 b,i). The marked sharpening of the corresponding band in the alkali treated outer layer spectrum compared with the untreated (Figure 1 b,o; 1 a,o) strongly suggests the presence of acetyl groups in the outer layers also.   The only other obvious effect of alkali treatment was to partially dissolve SiO_2, absorption of Si-0 in Figure 1 b,o being scarcely detectable.

The ATR spectra of the rumen treated straw (Figure 1 c) show significant differences from those of the original untreated straw.   At the inner surface, lignin was enhanced relative to polysaccharide by up to 10-fold, and acetyl groups were similarly concentrated (Figure 1 a,i; c,i).   This must have occurred as a result of degradation of both cellulose and hemicellulose in the inner surface layers because the general polysaccharide absorption pattern over the 1200-300

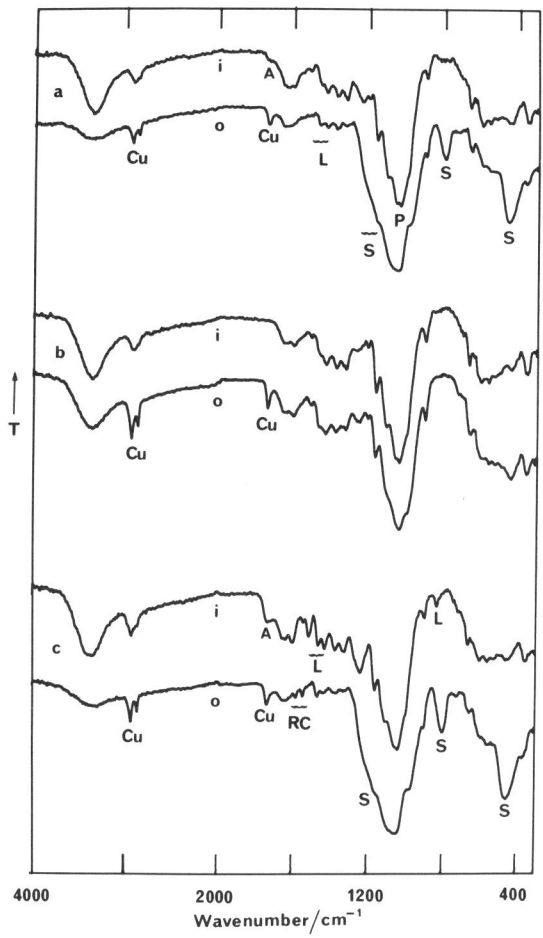

Figure 1. Attenuated total reflectance (ATR) infrared spectra of whole wheat straw (*Triticum aestivium*, cv Timor): a, untreated; b, alikali treated; c, after rumen digestion for 120 h. A, acetyl; L, lignin; P, polysaccharide; S, silica; Cu, cuticle; Cs, cellulose; RC, rumen components; i, inner straw surface; o, outer; T, % transmission.

cm^{-1} range was not significantly different from that of the original inner surface (Figure 1 a,i; c,i). In contrast, there were only relatively minor changes in the ATR spectra of the outer surface layer. Absorption arising from cuticle, lignin, polysaccharide and silica were virtually unchanged compared to the originals (Figure 1 a,o; c,o) indicating little or no attack on these components in the rumen. Weak additional absorption bands at 1580 and 1540 cm^{-1} and enhanced C-H absorption at 2920 and 2850 cm^{-1} especially in the spectrum of the outer surface RC in Figure 1c,o) are associated with rumen components which have spectral characteristics of polysaccharide possibly of bacterial origin.

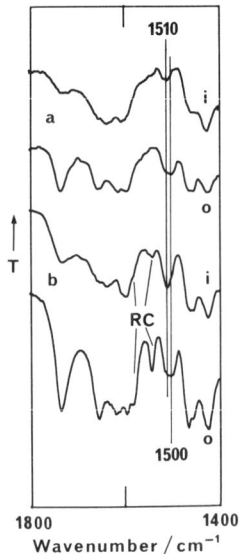

Figure 2. ATR infrared spectra of whole straw: a, untreated; b, after rumen digestion; i, inner straw surface; o, outer. Symbols as in Figure 1.

As noted previously, the lignin in the outer layer showed no change in spectrum after rumen digestion. This can be seen more clearly in Figure 2 a,o; b,o. That in the inner layer however underwent a change resulting in the original broad aromatic ring band at 1510 cm^{-1} becoming much sharper (Figure 2 a,i; b,i). Close similarity between this spectrum of modified lignin and that of extracted dioxan lignin suggests that the lignin accumulating at the inner surface of the rumen-incubated straw, like extracted lignin, has been at least partly released from bonding to other cell wall components.

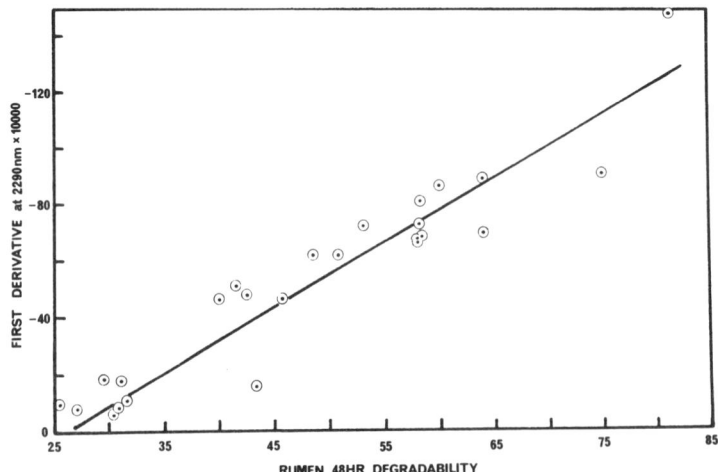

Figure 3. Plot of the first derivative optical data for 25 cereal straws at 2290 nm versus their 48 hour rumen degradability ($R^2 = 0.89$; SEC = 4.9). DEGRADABILITY = 26.1 - 4181 * 2290 nm.

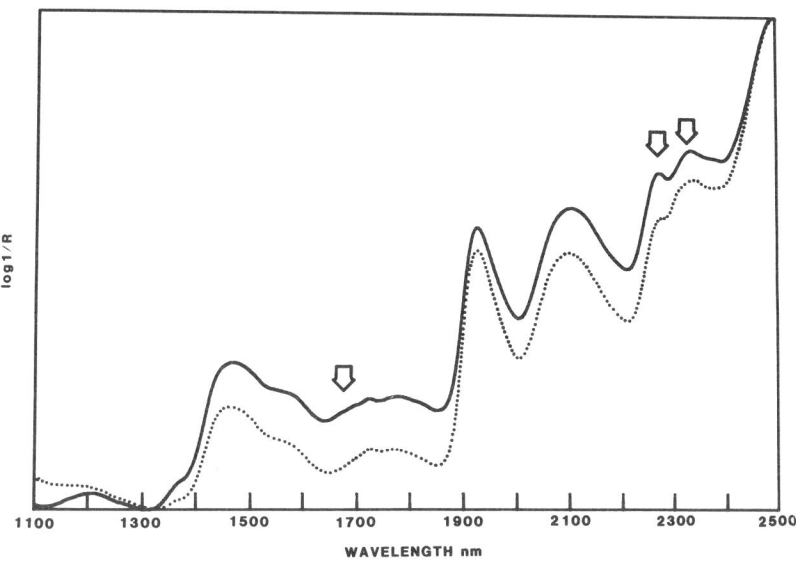

Figure 4. NIR spectra of cereal straw stem fraction (solid line) versus leaf fractions (broken line). Arrows show regions of shape difference.

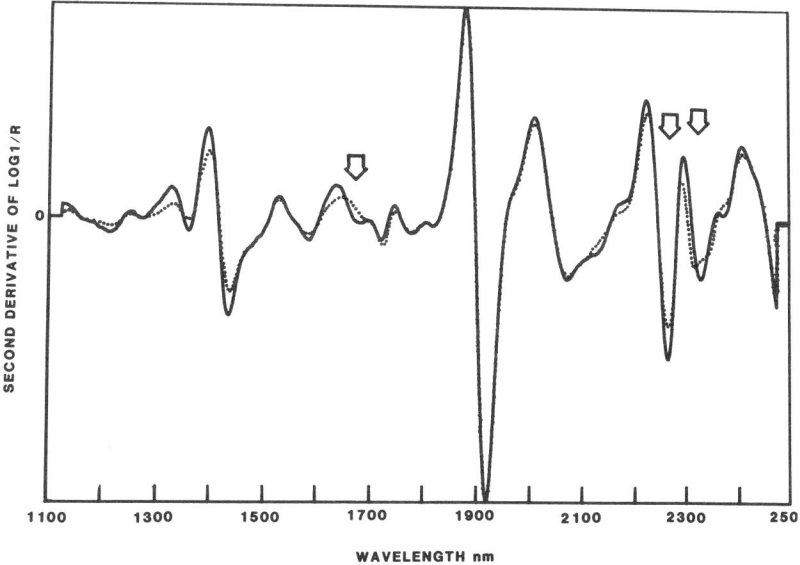

Figure 5. Second derivative NIR spectra of cereal straw stem fraction (solid line) versus leaf fraction (broken line). Arrows show main regions of shape difference

Preliminary results from a related study of oil-seed rape straw has also demonstrated the ability of ATR to distinguish inner and outer surfaces, and to identify changes occurring as a result of rumen incubation.

NIR.   Table 2 lists the NIR calibration equations for 48 hour degradability developed on all 25 straw samples (Table 1).   NIR spectra could explain 90% of the variance in degradability consistent with a standard error of 5%.   The most relevant wavelength was 2290 nm (first derivative) or 2310 nm (second derivative).   The linear plot of optical data at 2290 nm versus rumen degradability is shown in Figure 3.

When the mean spectrum (n=4) of leaf fractions of wheat straw is compared with the mean spectrum of stem fractions, three regions are found to be slightly different (Figure 4).   Their second derivative (Figure 5) accentuates these small differences in shape over the 1600-1700 nm and 2200-2300 nm regions.   When the correlation graph to degradability is computed using second derivative spectra of the 25 straws, it is clear that several regions are correlated (Figure 6). In particular 1676 nm, 2262 nm and 2310 nm are strongly correlated and have relatively large variances across the 25 samples.   The choice of NIR wavelength for measurement is a compromise between the best correlation and the highest standard deviation across the sample set.   The partial regression coefficients of these three bands are shown in Table 2.   Note that some pairs offer an improvement over single wavelengths whereas others do not.

TABLE 2.   NIR calibration equations for cereal straw dry matter degradability in rumen incubation (48 h)[1]

Statistics		Regression coefficients				Wavelength segments		
SEC	R2	B_0	B_1	B_2	B_3	1	2	3
						(First derivative spectra)		
4.99	0.89	26	-4181	-	-	2290'		
4.21	0.93	23	-8775	5571	-	1778'	2290'	
						(Second derivative spectra)		
5.74	0.86	36	6200	-	-	1676"		
8.55	0.69	123	2101	-	-	2262"		
5.63	0.87	28	-2340	-	-	2310"		
5.59	0.87	31	-1336	2725	-	2310"	1676"	
4.69	0.91	68	4578	821	-	1676"	2262"	
4.93	0.90	59	732	-1775	-	2262"	2310"	
4.76	0.90	65	-529	3305	769	2310"	1676"	2262"

[1] Data based on 25 samples with a mean 48 h degradability of 48.22% (SD, 15.4)

The 1676 and 2262 nm regions are already well established *"in vivo"*. DOMD estimates in a large set of silages[5].   These bands are inversely related to degradability and arise from lignin chromophores whereas 2310 nm is directly related to degradability.   This 2310 nm band has been ascribed to the methylene

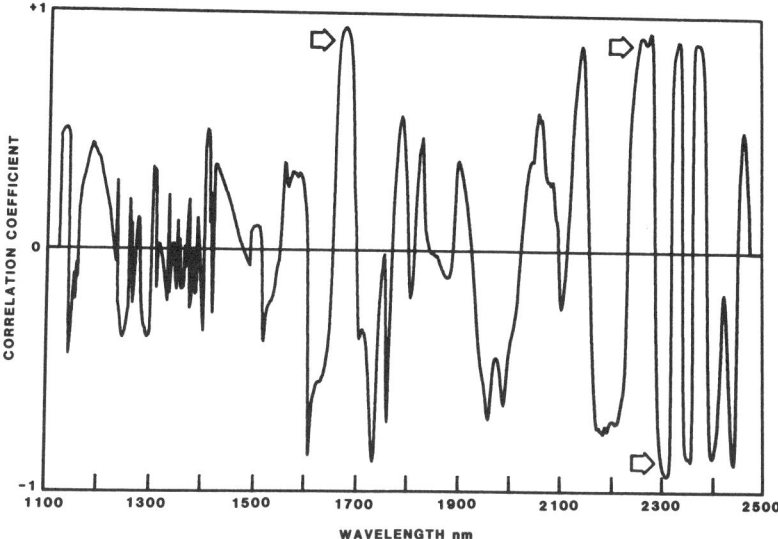

Figure 6. Graph of the correlation between the second derivative NIR spectra of 25 cereal straws and their 48 hour rumen degradability. Note that absorbances correlating positively are in fact indigestible components. Arrows indicate the most useful bands at 1676, 2262 and 2310 nm.

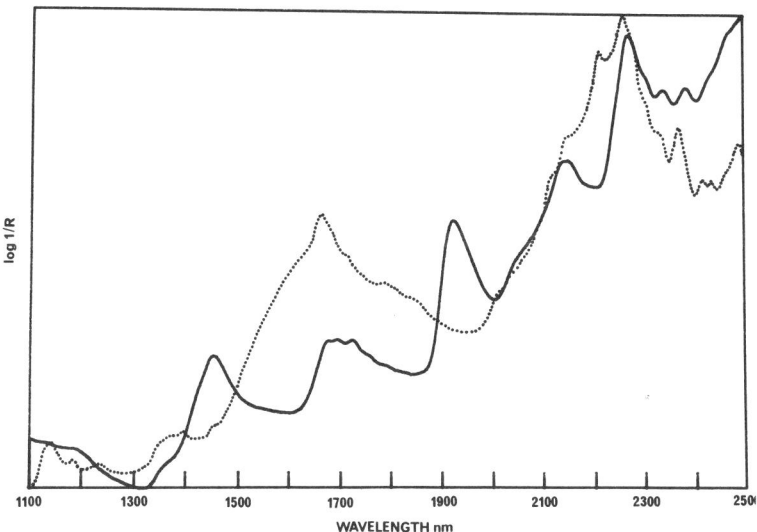

Figure 7. NIR spectra of chemically extracted straw lignin (solid line) and vanillin (3-methoxy-4-hydroxybenzaldehyde) (broken line). The 1668 nm and 2270 nm peaks are common to both.

stretch-bend combination (2850 + 1480 = 4330 cm^{-1}). The 1676 nm band is the aromatic CH stretch first overtone characteristic of phenolic compounds and lignin.[6]

The NIR spectrum of chemically extracted straw lignin is shown in Figure 7 which demonstrates absorption maxima at 1668, 2150 and 2270 nm (1440 and 1940 nm are water). This can be compared with the spectrum of vanillin (3-methoxy-4-hydroxybenzaldehyde). The 2270 nm band is a CH combination observed in alkylbenzene compounds (2270 nm = 4405 cm^{-1} = 1500 + 2905 cm^{-1}) arising from C=C and alkyl C-H substituents.

Any of these bands may be analytically useful for assessing straw degradability either alone or taken in pairs. However the number of samples used in this study may not be enough to ensure a robust NIR calibration.

These NIR results suggest a strong negative correlation between lignin content of ceral straw and its 48 h rumen degradability. This conclusion was reinforced by a similar correlation ($R^2 = 0.84$) between the MIR lignin aromatic

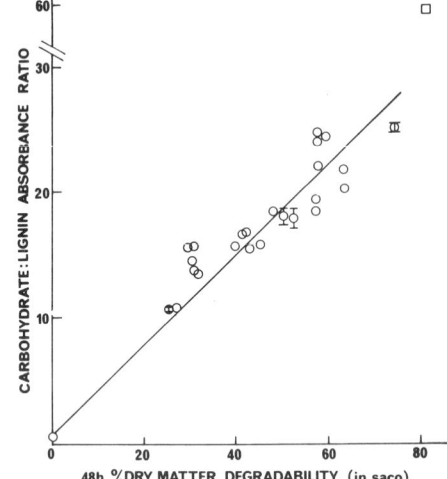

Figure 8. MIR calibration plot for 48 h dry matter degradability of straw and straw fractions, and lignin using carbohydrate (1050 cm^{-1}); lignin (1510 cm^{-1}) absorbance ratio, □, leaf blade sample (No. 1, Table 1) not used in correlation. I, error bars on some data for illustration.

C=C absorption band at 1510 cm^{-1} in normal transmission spectra and 48 h DMD for 24 out of the 25 samples plus an extracted dioxan lignin which was assumed to have a 48 h DMD of 0% (Figure 8). The sample which did not fit this correlation was the only leaf blade sample in the set (No. 1, Table 1) and had the highest degradability. The aromatic C=C band in the MIR spectrum of this sample (Figure 9a) was anomalous in both position and intensity relative to polysaccharide; compared to that for stem (Figure 9b) and all other samples, it had a higher frequency, 1516 cm^{-1} compared to 1510 cm^{-1}, had less than half of the relative absorbance, and was much broader. The implication is that the aromatics present in this leaf sample, which is enriched in cuticular waxes (stronger CH$_2$ stretch at 2920, 2850 cm^{-1}, Figure 9a) compared to the other samples, are not constituted as lignin *per se*, but perhaps as lignin precursors.

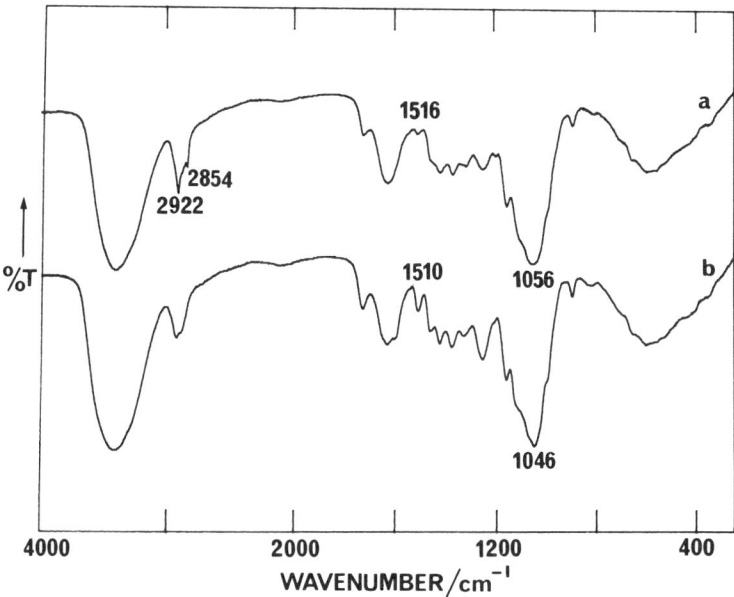

Figure 9. MIR infrared absorption spectra of barley straw: a, leaf blade (No. 1, Table 1); b, stem (No. 3, Table 1).

CONCLUSIONS

Reflectance spectra in the MIR and NIR regions are analytically useful in distinctly different aspects of cereal straw composition and degradation. MIR offers the advantage of more certain assignment of spectral bands to identify functional groups in chemical components during degradation, whereas NIR employs correlation transform spectroscopy to allow ease of sample presentation and measurement. Both NIR and MIR strongly support the involvement of bands derived from lignin as the most important feature correlating to rumen degradability of lignocellulose.

REFERENCES

1. Chesson, A. (1981). Effects of sodium hydroxide on cereal straws in relation to the enhanced degradation of structural polysaccharides by rumen micro-organisms. *Journal of the Science of Food and Agriculture* **32:** 745-758.

2. Adams, M.J. and Black, I. (1983). Applications of a microcomputer system in an infrared laboratory. *Journal of Automatic Chemistry* **5:** 9-13.

3. Shenk, J.S.; Landa, I,; Hoover, M.R. and Westerhaus, M.O. (1981). Description and evaluation of a near infra-red reflectance spectrocomputer for forage and grain analysis. *Crop Science* **21:** 355-358.

4. Russell, J.D.; Fraser, A.R.; Gordon, A.H. and Chesson, A. (1988). Rumen digestion of untreated and alkali-treated cereal straws: a study by multiple internal reflectance in frared spectroscopy. *Journal of the Science of Food and Agriculture* **45:** 95-107

5. Barber, G.D.; Givens, D.I.; Kridis, M.S.; Offer, N.W. and Murray, I. (1989). Prediction of the organic matter digestibility of grass silage *Animal Feed Science and Tecnhology* (In press).

6. Murray, I. (1987). The NIR spectra of homologous series of organic compounds. In: *Near Infra-Red Diffuse Reflectance Transmittance Spectroscopy*. (Hollo, J., Kaffka, K.J. and Gonczy, J.L. (eds.). Akademiai Kiado, Budapest.

7. Murray, I. and Williams, P.C. (1987). *ChemicalPprinciples of Near Infra-red Technology*. Williams, P.C. and Norris, K.H. (eds.). American Association of Cereal Chemists, St Paul Mn, U.S.A.

AN INTRODUCTION TO PYROLYSIS MASS SPECTROMETRY OF LIGNOCELLULOSIC MATERIAL: CASE STUDIES ON BARLEY STRAW, CORN STEM AND *AGROPYRON*

J.J. BOON

Mass spectrometry of Macromolecular Systems, FOM Institute for Atomic and Molecular Physics, Kruislaan 407, 1098 SJ Amsterdam, Netherlands

SUMMARY

Some basic principles of mass spectrometry and an introduction to the pyrolysis MS and GCMS of lignified plant cell wall material are presented. PYMS visualises molecular weight distributions of the thermal dissociation products of thermally activated polymer systems. PYGCMS gives compound distribution profiles which are used for structural identification of specific marker compounds. Analysis of the MS data by multivariate analytical techniques yields correlation patterns in which discriminating sets of mass peaks can be traced and related to structural changes. The relative differences between samples are expressed in a geometric format as graphs and maps.

Analytical pyrolysis data on Barley straw are used to illustrate PYMS and PYGCMS. Unique higher mass data from PYMS under chemical ionization conditions in the ion source point to dimers and trimers released from the polysaccharides and lignin. PYMS mapping by multivariate analysis is shown in a comparative study on corn stem from *Zea mays* L.var. BM, Eta Ipho and LG11 and on several *Agropyron* species and hybrids illustrating the potential for rapid characterization of plant varieties. PYMS of micromanipulated samples from parenchyma and vascular bundles of corn stem demonstrates the potential for detailed studies on a microscopic level.

SOME PRINCIPLES OF MASS SPECTROMETRY

Mass spectrometry is, in principle, a destructive analytical technique in which chemical compounds are identified by analysis of their molecular ion and fragment ions *in vacuo*. These ions are separated by electric and magnetic fields yielding a distribution of the mass/intensity of the usually monovalent ions. A MS system consists schematically of:

INLET*ION SOURCE*E.M.SEPARATION*DETECTOR--DATA SYSTEM

The inlet can be a direct probe or a chromatographic inlet, ions can be made by electron impact ionization (E.I.), chemical ionization (C.I.), photoionization (P.I.), field Ionization (F.I.) and other less well known methods. The separation of the ions occurs in electromagnetic fields such as a quadrupole Rf field or by a combined electrostatic analyser and a magnetic field in double focussing instruments (E/B). Mass separation is achieved by scanning of the electromagnetic field past an entrance slit to the electron multiplier of the detector. Most routine instruments are nowadays equipped with computers for data acquisition and processing. Detailed descriptions can be found in the literature[1,2]. In general a mass spectrum shows the molecular ion of the compound and smaller fragment ions resulting from dissociation of overactivated molecular ions in the ion source due to the electron bombardment process to produce the ions. Examples of the spectrum of sinapyl aldehyde and 1,6-anhydroglucopyranose (levoglucosan) obtained at 70 eV taken from Barley straw PYGCMS (pyrolysis gas chromatography mass spectrometry) data are given in Figure 1

The synapyl aldehyde spectrum shows a strong molecular ion at m/z 208 and a few relative small fragment ions due to the stability of this aromatic compound. The levoglucosan spectrum does not show the molecular ion (M^+=162) at all but shows instead strong fragmentation ions at m/z 57, 60 and 73 due to the instability of the molecule (the molecular ion can be obtained by CI). Much effort is devoted by mass spectrometrists to relate the distribution pattern of the ions with the molecular structure of the molecule[3]. The dissociation of molecules is determined by the molecular structure and the internal energy distribution upon ionization. In the last decades several so-called soft ionization techniques such as field ionization (ionization by tunnelling at high voltage on a surface) and chemical ionization (ionization by charge exchange with a partially ionized gas such as methane or ammonia) have been developed on commercially available instruments which minimize this internal energy and hence produce spectra with only the molecular ion or a cluster of the reaction gas with the molecular ion. Measurement of the molecular ion at high resolution in a double focussing E/B sector mass spectrometer will yield its elementary composition. Such information combined with the distribution and elementary composition of the fragment ions obtained by electron impact ionization is often enough to postulate a chemical structure or to prove a structure by comparison of the spectrum with a standard compound. The distributions of the fragment ions obtained by electron impact ionization at 70 eV have been found sufficiently stable to compile atlases of mass spectra obtained at nominal resolution[4] which can be used for routine identification of gas phase compounds.

The mode of ion formation is revolutionized in recent years by sputtering from highly viscous liquid phases in the ion source. Traditionally gas phase apolar molecules were admitted by a heatable probe or through a gas chromatographic inlet. Polar nonvolatile material and thermally labile compounds can now be admitted and ionized directly from sputter probes using fast atom bombardment (FAB) ionization or through liquid chromatographic inlets such as thermospray or the continuous flow fast atom bombardment probe. In these latter methods droplets are formed either by a nozzle or by sputtering upon impact with a beam of neutral atoms such as Argon or Xenon. Charge exchange phenomena in the liquid droplets lead to the formation of ions ready for analysis in the MS. These new ionization techniques have proved especially important for the analysis of biochemical compounds[6]. The MS analysis of the

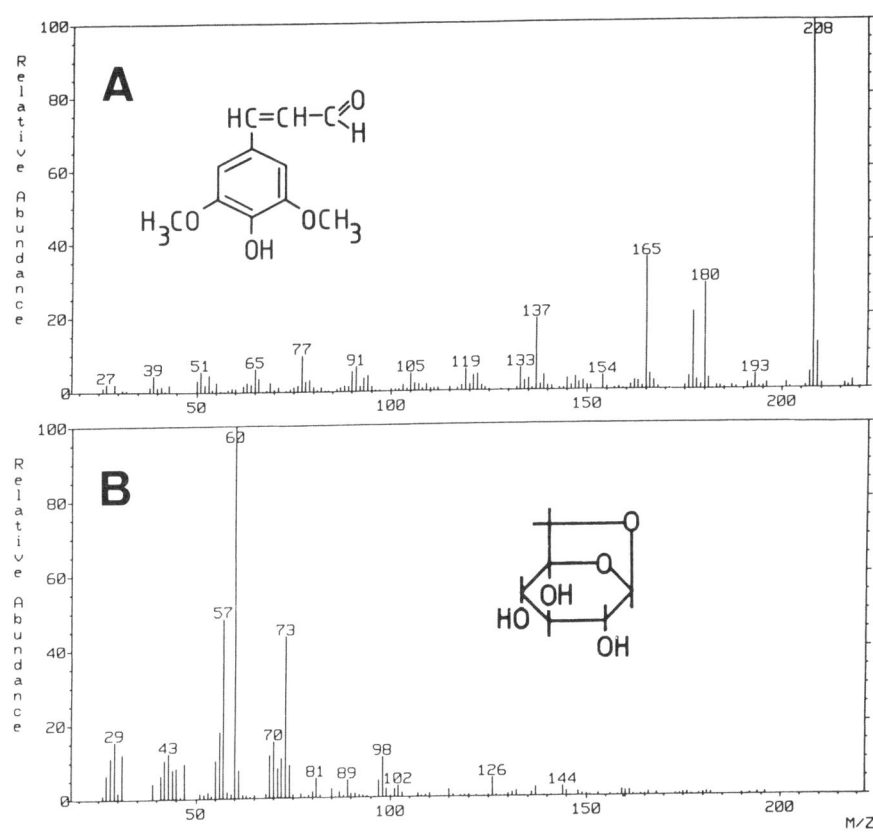

Figure 1. Mass spectra at 70 eV under electron impact ionization conditions from sinapylaldehyde (A) and 1,6-anhydroglucose (B).

medium molecular weight molecules (mass 1000 - 10,000 D) however is certainly not routine yet and requires expensive instrumentation and skilled technical staff.

More complex materials such as synthetic polymers, high molecular weight biomaterials such as plant cell wall material, whole bacteria, soil organic matter, etc. can not be directly analyzed. Thermal dissociation of these complex materials yields smaller fragments which fall within the range of present day MS systems. The discipline involved is called analytical pyrolysis and the instrumentation is usually indicated by the prefix pyrolysis (PY).

ANALYTICAL PYROLYSIS BY PYMS AND PYGC(MS)

Two methods of analytical flash pyrolysis are employed in our laboratory at FOM for PYMS studies i.e. Curie point pyrolysis using inductive heating of ferromagnetic probes and platinum filament pyrolysis using resistive heating. Both devices can be used in a PYMS and in a PYGCMS mode. Curie point pyrolysis MS is now employed in a variety of instruments modelled after the first PYMS system built at FOM by Meuzelaar, Kistemaker and coworkers in the early seventies[7]. The FOMautoPYMS - a low voltage EI quadrupole MS- has a fully automated sample exchange system and a transmission of pyrolysate comparable to what can be injected into the gas chromatograph with the FOM-3LX pyrolysis unit[8]. Low voltage E.I. at about 15 eV is employed to obtain molecular weight distributions of the pyrolysate with a minimum of fragmentation due to the ionization process. Commercially available PYMS systems are the VG Pyromass 8-80 and the Horizon PYMS-200X.

Figure 2 shows a schematic of the pyrolysis units for Curie point pyrolysis work. In the PYMS the pyrolysate is generated *in vacuo* by inductive heating of the ferromagnetic sample probe and the pyrolysate drifts via a heated gold coated expansion chamber to the ion source. The pyrolysis unit in PYGCMS is simply comparable to other injection system to a GCMS. The pyrolysis takes place in the carrier gas stream and volatile parts of the pyrolysate are injected into the GC. The great advantages of Curie point pyrolysis over other heatable sample probes are the flash heating conditions (about 8000 K/s *in vacuo*), the reproducibility of the heating process, the ferromagnetic probes in exchangeable glass liners, the disposability of the inexpensive sample probes (a fresh clean wire for each analysis). Sample exchange can be automated by exchange of the glass liners after the analysis. The high frequency inductive field can upset the electromagnetic fields in the ion source and the pyrolysis process has to take place for this reason at some distance from the analyser which hampers the transmission of higher molecular weight fractions. This problem can be overcome by platinum filament pyrolysis performed in the ion source, a method also known as D(irect) C(hemical) I(onization) MS when chemical ionization is used to produce the ions. Figure 2 shows a schematic of the experimental arrangement. Small amounts (a few micrograms) of liquid or solid sample are mounted on the sample loop which is inserted through a vacuum inlet into the source house. The platinum wire (diameter 100 micron) is heated by resistive heating with heating rates from 0.5 about 50 K/s to its final temperature. With this method we have observed molecular weights up to about mass 2000 in pyrolysates of some biomaterials because of the improved transmission to the mass spectrometer and the soft-ionization conditions. Naturally the analysis can also be performed under low voltage EI conditions for comparison with Curie point pyrolysis EI modes. Schulten *et al.*[9] have chosen for a micro-oven type of pyroloysis in which a glass tube with sample inserted in a heatable probe (max. temp. 1000 K, heating rate up to 10 K/s) protrudes in a field ionization ion source (F.I.). Interesting results on beech leaves, lignins and soils have been obtained in this way, showing that higher molecular weight materials (up to about m/z 650) are present in such pyrolysates. Although F.I. has great advantages such as minimal fragmentation, almost no formation of cluster ions, low compound selectivity and high sensitivity, the method is not widely used because it requires the more expensive E/B mass spectrometers and great skill to obtain good results.

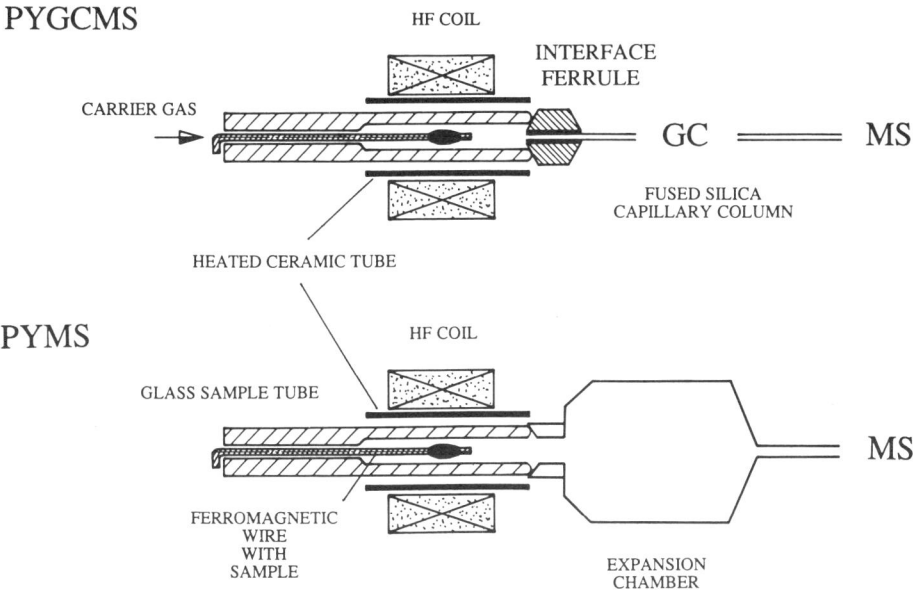

PYGCMS

HF COIL

INTERFACE
FERRULE

CARRIER GAS

GC ═══ MS

FUSED SILICA
CAPILLARY COLUMN

HEATED CERAMIC TUBE

PYMS

HF COIL

GLASS SAMPLE TUBE

MS

FERROMAGNETIC
WIRE
WITH
SAMPLE

EXPANSION
CHAMBER

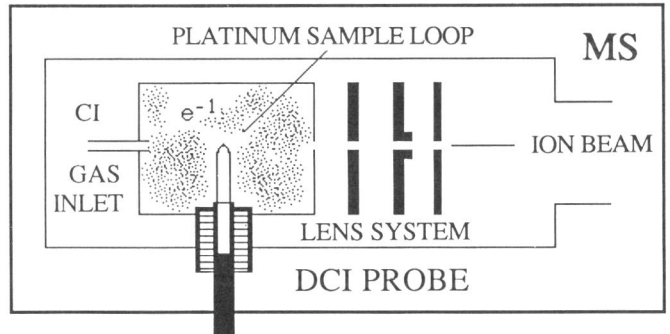

PLATINUM SAMPLE LOOP

MS

CI

e^{-1}

GAS
INLET

ION BEAM

LENS SYSTEM

DCI PROBE

Figure 2. Schematic diagram of the pyrolysis units for Curie point PYGCMS, Curie point PYMS and platinum filament pyrolysis in the ion source (C) under EI or CI conditions (DEI-MS and DCI-MS)

Pyrolysis-gas chromatography-mass spectrometry

The results in PYGCMS depend not only on the injection characteristics of the pyrolysis unit but especially on the chromatographic parameters chosen for separation of the pyrolysis products generally requiring capillary GC at best resolution. Both Curie point pyrolysis and platinum filament pyrolysis units (e.g.[12,29]) are available commercially. The FOM-3LX unit[8] is the only splitless injection system in which the glass liner is placed directly on a Kalrez interface with the capillary column inlet (see Figure 2). This system allows the analysis of very small amounts of sample and thin sample layers which are important variables from the point of view of reproducibility and control on internal energy of the molecular system during pyrolysis. The chromatographic problem at hand is the base line separation of a large number of compounds in a mixture with a high diversity in molecular weight (ranging from C_1 to C_{50}), in polarity due to O, N and S containing groups in aliphatic and heterocyclic compounds and in stability. Peak broadening due to poor injection characteristics of the pyrolyser can be partly overcome by cold trapping of the pyrolysate.

We have successfully used long apolar capillary column for mainly aliphatic pyrolysates[10] and apolar thick film (1 micron) capillary columns for analysis of lignocellulosic materials[8,11]. Others[12,13] have shown good results for polysaccharide and wood pyrolysates respectively on polar phases such as OV-1701, but in general the molecular weight range of pyrolysates of plant and bactertial material poses elution problems which can be partly overcome by the use of H_2 as carrier gas and the use of shorter columns. An alternative is off-line pyrolysis and derivatisation by methylation and silylating agents[14]. In the future HPLC and SFC will be more widely used for the analysis of the polar fractions. Off-line pyrolysis of cellulose and HPLC of the pyrolysate has shown the presence of cellobiosan (a β-1,4-glucose-[(1,6)-anhydro]-glucose)[15].

Sample preparation

Liquid samples are simply applied to the analysis probes by application with syringes. Solid samples are usually applied to the wires from suspension in water or methanol by calibrated pipets using a flamed glass capillary as an extension. The droplets of suspension are dried *in vacuo* in a specially designed vacuum chamber in which a number of sample wires can be prepared at the same time (see ref[7], p32). Sample amounts applied are somewhat dependent on the nature of the sample but should not be higher than about 50 micrograms. Solid samples are homogenized by grinding in liquid nitrogen cooled mills. Best results with woody samples have been obtained by filing slowly with metal file[16]. Coal samples have been applied by sampling from a slurry[17]. A pressing method was developed for synthetic polymers[18].

ANALYTICAL STRATEGY IN FINGERPRINTING AND STRUCTURAL STUDIES

Pyrolysis mass spectrometry is a relatively rapid method for obtaining fingerprinting information on a molecular level. The method has been applied to a wide range of organic materials from samples of living plants, bacteria, fungi, animal and human cells and their decayed remnants, to environmental samples such as soils, particulate organic matter in natural waters, marine environments, polluted sites, and also geological samples and meteorites. Low voltage EI PYMS spectra of a wide range of materials have been published as an atlas[7].

PYMS data is time integrated information, although a time resolved acquisition of the data is possible in principle. In practice spectra are signal averaged to one spectrum or summarized afterwards. There are several problems in the interpretation of PYMS data. First, the problem of the identity of the individual mass peaks which is complicated by the presence of isobaric compounds and by isomeric molecular structures. The presence of isobaric masses with a different chemical composition can be shown in principle by measurement of the elementary composition under HR(high resolution)MS conditions but there are serious problems with sensitivity. Secondly, the relation of a mass peak distribution in a spectrum i.e. the individual pyrolysis products, with the structure of the polymer (nature of the constituent, their relative distribution, their sequence). Detailed analysis of well characterized standards of biopolymers help to understand this relationship. Thirdly, the relative distribution of the mass peaks, identified as representative of a polymeric constituent, and their relationship with the actual concentration in the sample analysed. So far most of the attention in the field has been given to solve the first two problems mentioned.

Analysis of the pyrolysate by PYGCMS using comparable pyrolysis conditions is a useful approach to identify pyrolysis products and to correlate mass peak information with molecular structure. Figure 3 shows a diagram of a gcms data file which can be read as mass spectrum/unit of time with individual peaks in the total ion current trace but the time axis can be traced as well along one mass value yielding a mass chromatogram. Summation of part or all of the ms data in the gcms file will yield a summary spectrum which can be compared with PYMS data obtained without GC separation. In this way specific mass peaks can be traced back to the full mass spectrum and its corresponding pyrolyis product identified. The correlation of mass spectrum of a compound and its molecular structure is a general task in mass spectrometry, but lately GC-FTIR in combination with MS has improved the possibilities.

PYGCMS data can be used directly for profiling of marker compounds. The method is relatively time consuming because there is still a large portion of interaction required with the mass spectrometrist to assign the compounds. If a molecular system under analysis is well known, the time required to process the data file can be cut back to critical questioning of the data system. Analysis time depends on the chromatographic conditions but generally requires about an hour for a complete pyrolysis product profile under high resolution GC conditions.

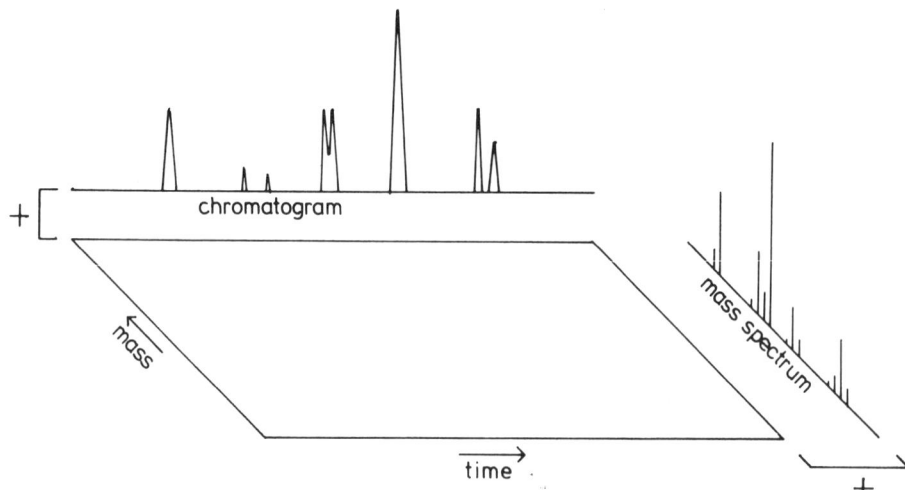

Figure 3. Schematic of a GCMS data matrix. Post-analysis single mass monitoring along the time axis gives mass chromatograms (see Figure 10). Time integration of the data leads to summary spectra (see Figure 9).

Thus in mapping of complex biomaterials, PYMS, PYGC and PYGCMS are used interactively in an analytical strategy as shown schematically in Figure 4. PYMS is used for rapid screening of a sample set with multivariate analysis as a profiling and mapping technique. Preliminary identifications are confirmed by PYGCMS. Pyrolysis conditions and chromatographic conditions are best studied in PYGC mode before more detailed PYGCMS analysis.

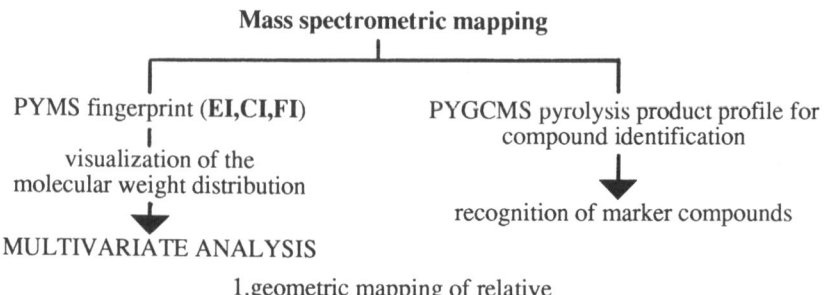

Figure 4. Scheme for the mapping of complex biomaterials

Multivariate analysis of pyms data

Especially in automated PYMS systems large sample sets can be generated in a short time which made the development of multivariate data analysis methods for the mass spectra a necessity[19]. Factor and discriminant analysis of the mass spectra is used compare the spectra and to obtain a qualitative and quantitative view of the differences between them[20]. Spectra are considered to be points in a multidimensional space with the mass numbers as coordinate axes. The relative distribution of the mass intensities in each spectrum determines the position in the PYMS universe. Similar spectra will cluster as one group. Mathematically, the differences between the individual spectra are determined by comparison with the overall average spectrum. These data are factor analyzed to produce sets of correlated mass peaks (factors). If multiple analyses are available discriminant analysis can be performed which takes the reproducibility of the analytical data into account. The sets of correlated masses in factors and discriminant functions are often found to correlate with the mass peak distribution of isolated polymers. This can be understood because a polymer will often generate a number of pyrolysis products with different mass. Variation of such a polymer in a set of complex materials will lead to highly correlated variations in distribution of certain mass peaks characteristic for the polymers. Quantization leads to a description of the relative differences in the data set by socalled factor or discriminant scores which can be expressed in graphs, maps and 3D representations. Absolute quantization is possible by comparison with calibration curves obtained in the same way. The multivariate techniques have shown the remarkable reproducibility of pyrolysis fingerprinting and thus of the pyrolysis mechanisms involved. This stability is the basis for its usefulness in the characterisation of samples of (agro)biological importance.[16,21-24,30,32,37]

LIGNOCELLULOSIC MATERIALS

OECD Barley straw - a comparison of PYMS and PYGCMS data

PYMS fingerprinting. Pyrolysis analytical data on OECD barley straw cell walls will serve as a case study to show the potential of low voltage EI PYMS, chemical ionization PYMS and PYGCMS analytical data for agrobiology. A detailed account of barley straw treated by enzymes and wet chemical methods will appear elsewhere (Boon and Hartley, manuscript in prep.). The barley sample was a gift of R. Hartley from A.G.R.I. and has been subject of PYMS and PYGCMS studies before[23,24]. The sugar analysis of this material yielded 2.0% arabinose, 21.6% xylose, 47.2% glucose and 2% uronic acids. The total amount of phenolic acid was 8.6 mg/g with 5.7 mg/g *trans* p-coumaric acid and 2.2 mg/g *trans* ferulic acid (data supplied by Hartley).

Figure 5 A and 6 A shows the PYMS spectrum of OECD Barley straw obtained under low voltage EI conditions in the FOMautoPYMS and by DEI-MS on the JEOL DX303 E/B MS. The FOMautoPYMS spectrum shows data up

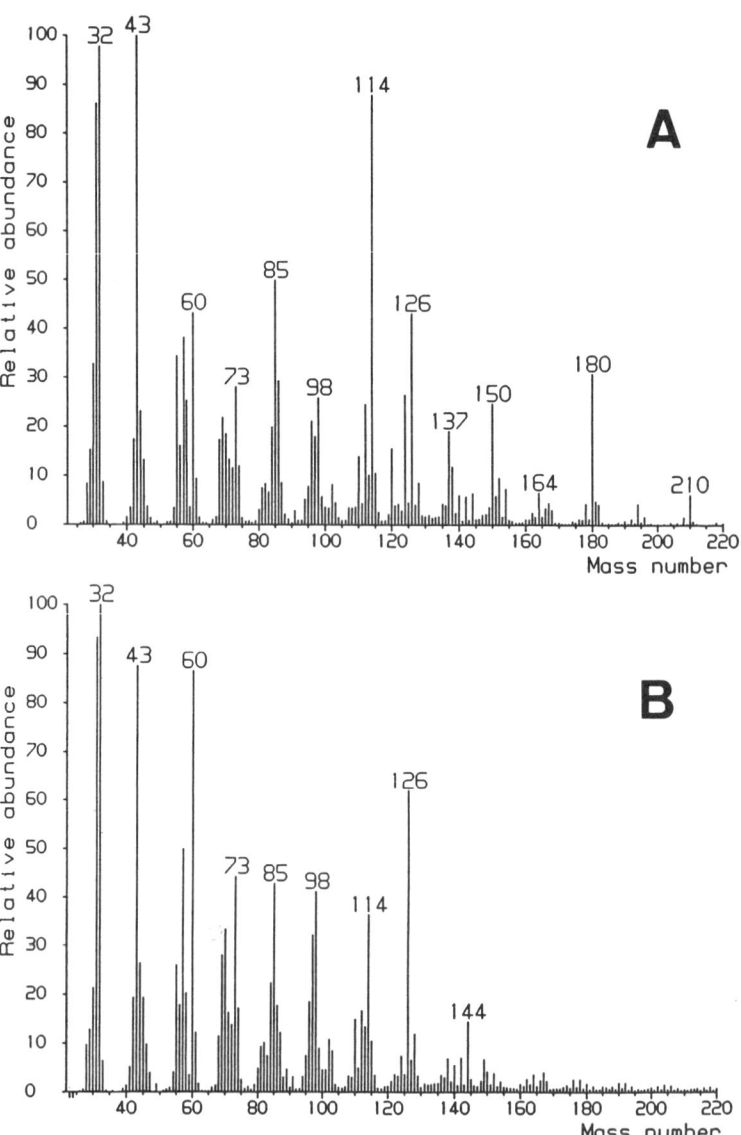

Figure 5. Low voltage EI PYMS spectra for native OECD barley straw (A) and microcrystalline cellulose (B) obtained on the FOMautoPYMS. Samples were applied from a water suspension and analysed at 510°C (total pyrolysis time 0.8 s) with a heated pyrolysis chamber.

to mass 220, the mass limit of the system. The DEI-MS spectrum was scanned up to mass m/z 1000 at a scan speed of 1 s/cycle and shows mass data up to m/z 450. The similarity of the two spectra in the lower mass range is remarkably good which indicates that Curie point pyrolysis and platinum filament pyrolysis yield similar pyrolytic fragments despite differences in heating rate and instrumental conditions. In general, the mass peak information below m/z 118 in low voltage EI spectra of lignocellulosic material is mainly from pyrolysis products of polysaccharides, where many of the mass peaks above m/z 118 with the exception of m/z 126, 128, 144 are from lignin derived phenolic compounds (see also[11]). As demonstrated in Figure 1 and Figure 5 B - a low voltage EI PYMS of cellulose-, the EI conditions are detrimental for the less stable polysaccharide marker compounds yielding m/z 57, 60, 73 and other low mass ions in the PYMS spectrum[26]. The pattern of m/z 120, 124, 137, 138, 150, 152, 154, 164, 166, 168, 178, 180, 182, 194, 196, 208, 210 is typical for woody angiosperms.[8,13,16,26] Monocotyledonous plants usually have relatively high m/z 120 and 150 from p-coumaric and ferulic acids either present ester bound to the polysaccharide fractions or ether bound in the true lignin structure. The mass peaks from m/z 250 and higher at relatively low intensity are presumably from di and trimeric constituents released from lignin. The m/z 328, 358, 388 and 418 suggest a homologous series of dimers such as desmethoxypinoresorcinol, pinoresorcinol (dimer of coniferyl alcohol), medioresinol and syringaresinol (a dimer of sinapyl alcohol) respectively. Some of these mass peaks have been observed also in a PYFIMS of Braun's lignin prepared from wheat straw[27].

Pyrolysis under CI conditions using ammonia as the reaction gas brings the polysaccharide fraction to attention. The spectrum of barley straw and of cellulose (for comparison) is shown in Figure 6 B and C. The spectrum of cellulose is dominated by m/z 180, a cluster ion of ammonia with the molecular ion of levoglucosan, an important pyrolysis product of cellulose. The m/z at 162, 134 and 144 correspond to m/z 144, 126 and 116 in the EI data. The compounds corresponding to m/z 116 however is relatively unstable and fragments further under EI conditions (unpublished GCMS data,[29]). The m/z at 342 and 504 point to $[(\beta\text{-}1,4)\text{-glucose}]_n$-anhydroglucose dimers and trimers. Comparable ions have been observed in laser desorption FTICR-MS of cellulose[28]. The spectrum of Barley straw in Figure 6 B shows a number of the cellulose mass peaks at m/z 134, 144, 162, 180, 222, 342, 402 and 504. The high EI peak at m/z 114 from dianhydropentoses in Figure 6 A is also the highest ammonia cluster ion at m/z 132. Comparative studies with xylan and arabans under similar conditions also show m/z 150 (anhydropentose), which is present in the Barley spectrum as well. Higher mass peaks from pentoses are m/z 264 and 282, from dimers and m/z 414 from trimers.

The phenolic compounds react differently under ammonia CI conditions because their mode of ion formation is strongly dependent on the molecular structure of the side chain at C_4 (the reactivity of the various phenols under CI conditions is the subject of further investigations). Phenolic acids in ethyl acetate extracts of the basic hydrolysate show ammonia cluster ions for p-coumaric acid ($M^+=164$) at m/z 182 and for ferulic acid ($M^+=194$) at m/z 212

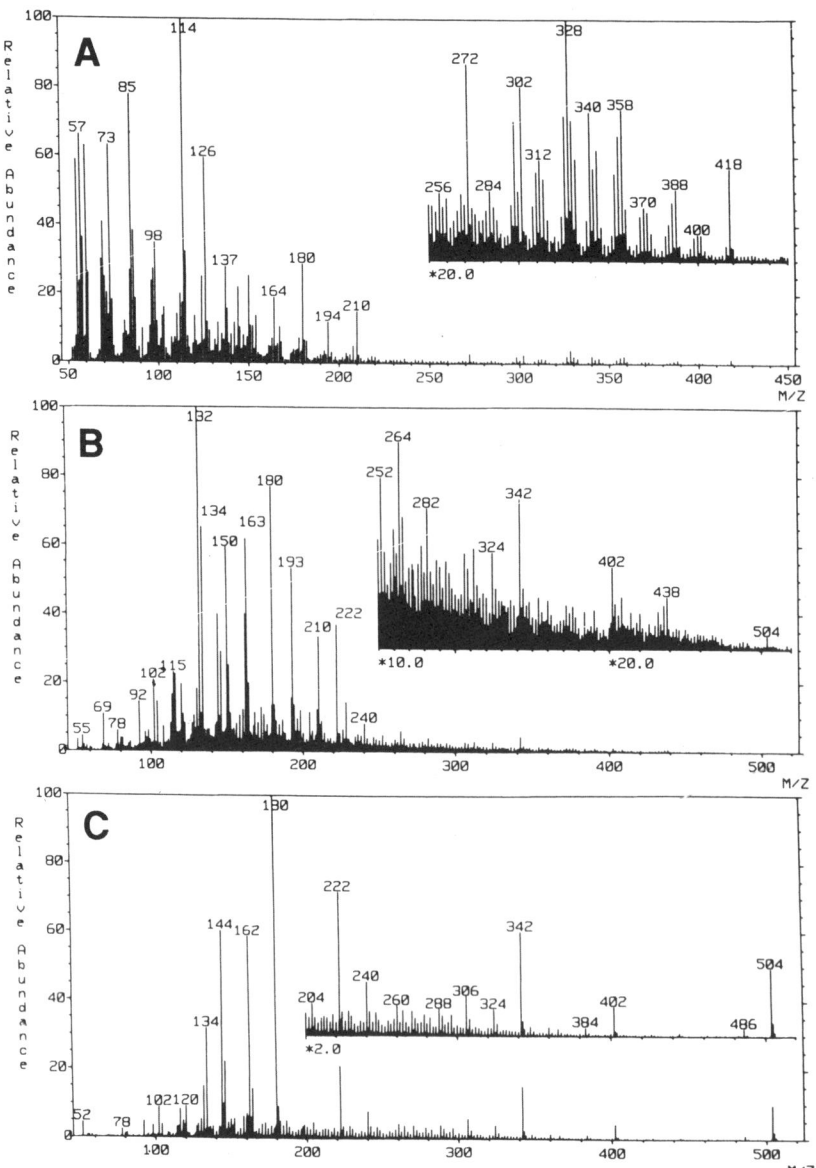

Figure 6. Platinum filament pyrolysis spectra of native OECD barley straw analysed under low voltage EI (about 17 eV) conditions (A) and ammonia CI conditions (B) and of microcrystalline cellulose under ammonia CI conditions (C). Data were obtained with the JEOL DX-303 double focussing MS (mass range m/z 20/1000; scan speed 1 sec/cycle; data acquisition time 60 seconds; all spectra summarized).

(unpublished results). These ions are relatively low in Figure 6 B, because during thermal dissociation the process of decarboxylation is more important than deesterification of the phenolic acid esters. The resulting vinylphenol and methoxyphenol show M^+ ions at m/z 120 and 150 in low eV EI (see Figure 6A) and the M^+ and M^+H ions at 121 and 151 under NH_3-CI. The phenolic compounds with a propenylalcohol side chain such as coniferyl alcohol and sinapyl alcohol show m/z 163 and 193 respectively pointing to a complex reaction in the ionization gas. Many of the other phenolic compounds do not show a clustering reaction with ammonia and simply have similar molecular ions as under EI conditions. The spectrum above m/z 250 appears to be dominated by polysaccharide pyrolysis product ions. Thus EI conditions are more efficient for profiling of the lignin derived pyrolysis products, whereas the higher molecular weight polysaccharide pyrolysis products are clearly dissociated further. These labile polysaccharide pyrolysis products are better visualized under ammonia CI conditions.

OECD BARLEY STRAW PYGCMS data

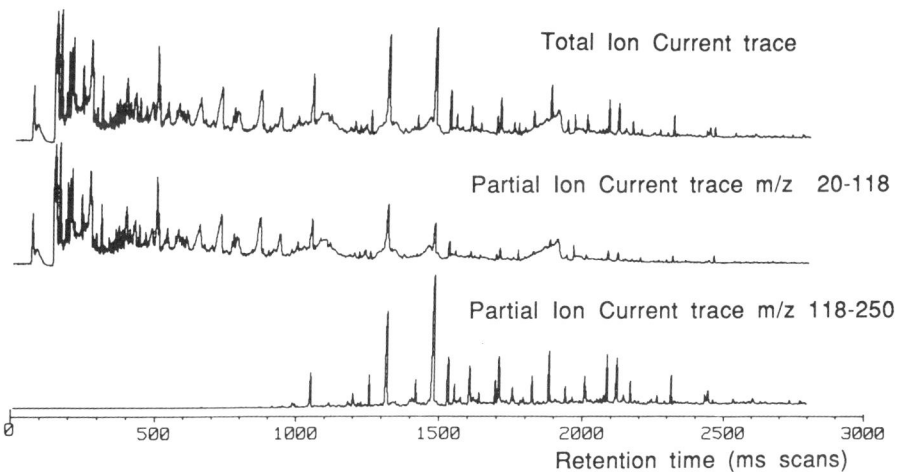

Figure 7. Total ion current and partial ion current traces of OECD barley straw PYGCMS data obtained by gas chromatography in helium on a 50 m fused silica capillary column (diam .32 mm) coated with CP-SIL 5 (film 1 μ) and MS analysis in the JEOL DX-300 (direct coupling, 70 eV EI).

PYGCMS profiling. The barley straw sample was also analyzed by PYGCMS under 70 eV ionization conditions on a thick film (1 μm) apolar fused silica capillary column. Figure 7 shows the total ion current (TIC) profile, a partial TIC from m/z 20-118 and a partial TIC from 118-250 referred to as ligninogram. The TIC on this apolar phase shows many unresolved broad peaks and a large number of fronting peaks due to rapid overloading of highly polar compounds. Practically all these peaks are from pyrolysis products of polysaccharides and many of those are now identified[11,29]. Much of this information is in fact redundant and their molecular structure is relatively nonspecific although the relative distribution of compounds has a relationship with the linkage of the monomers in the polysaccharides[30]. Sharper peaks, observed at longer retention times are from pyrolysis products of lignin. The partial TIC's demonstrate this phenomenon: m/z 118-250 shows no peaks below scan 1000, whereas m/z 20-118 is dominant in the first 1000 scans and a few anhydrosugars eluting later. The trace from m/z 118-250 can be used to filter out most of the polysaccharide derived peaks in a summary mass chromatogram. Figure 8 shows the profile in more detail together with a list of identified compounds. Many of these compounds have been observed before in beech wood and beech wood derived polymers[8,11,31] and are evolved from a mixed guaiacyl-syringyl ether bound lignin. There is a predominance of syringyl derived compounds in the barley pyrolysate. The structure of the side chain at position 4 in the phenols is related to the polymeric structure of the lignin but the exact relationship with the dimeric and trimeric units of coniferyl and synapyl alcohol is presently unknown[26]. A special feature of barley straw is the relatively high abundance of vinylphenol (peak 3) and vinylguaiacol (peak 5) which are partly formed by decarboxylation from the p-coumaric and ferulic acid esters and ethers [m/z 120 is substantially reduced after basic hydrolysis of the straw]. The coniferyl and synapyl alcohol, which are relatively abundant in woody material, are absent in barley. A small amount of ferulic acid methyl ester (peak 31) was observed in the PYGCMS data as well as dimeric cinnamic acid ester (not shown) which are presumably free volatiles in the straw.

The molecular ions of the phenolic compounds are listed as well and can be retraced in the summary spectra of the PYGCMS data in Figure 9 and in Figure 5 A. Both spectra in Figure 9 are used to relate the mass peak distribution of the PYGCMS data with PYMS (Figures 5 and 6 A). These spectra corroborate the earlier statement that polysaccharide information is mainly bunched below m/z 118 in EI MS data whereas lignin derived peaks are prominent between m/z 118-250.

A few specific pyrolysis products were selectively traced in mass chromatograms for m/z 144, 126, 114, 85 and 73. Structures for some of the peaks and their id's are given in Figure 10. The m/z 85 trace shows a broad peak from a γ-lactone derivative generated from methylated uronic acid polymers[32]. This peak is of importance in the discrimination of ammonia treated and native barley straw[24]. The m/z 114 shows several isomers but the main peak is from the 3-hydroxy-2-penteno-1,5-lactone generated from xylans[33]. The m/z 126 shows up in a number of pyranone derivatives from cellulose[11,29] and

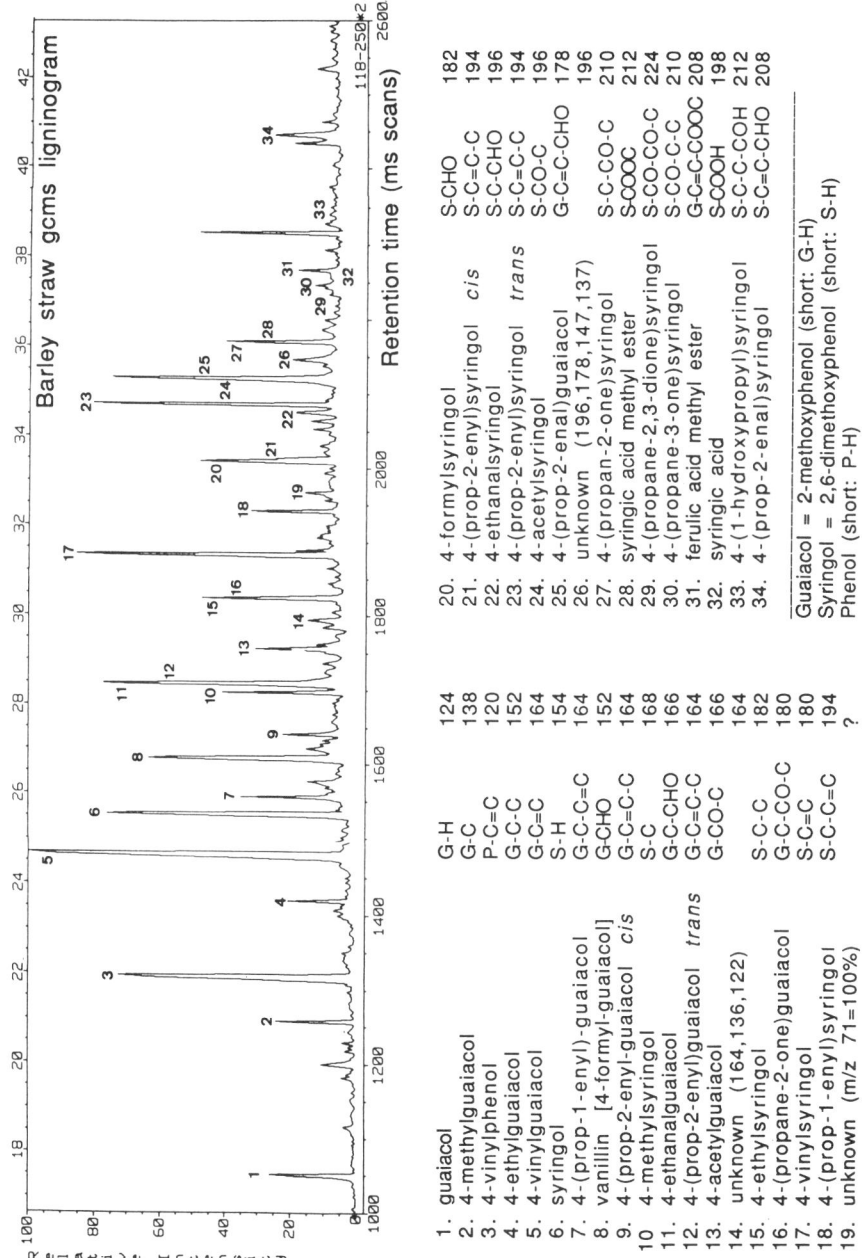

Figure 8. A "ligninogram" i.e. a partial ion current chromatogram from m/z 118-250 of OECD barley straw PYGCMS data (see Figure 7). The compounds were identified by their 70 eV EI mass spectra[8].

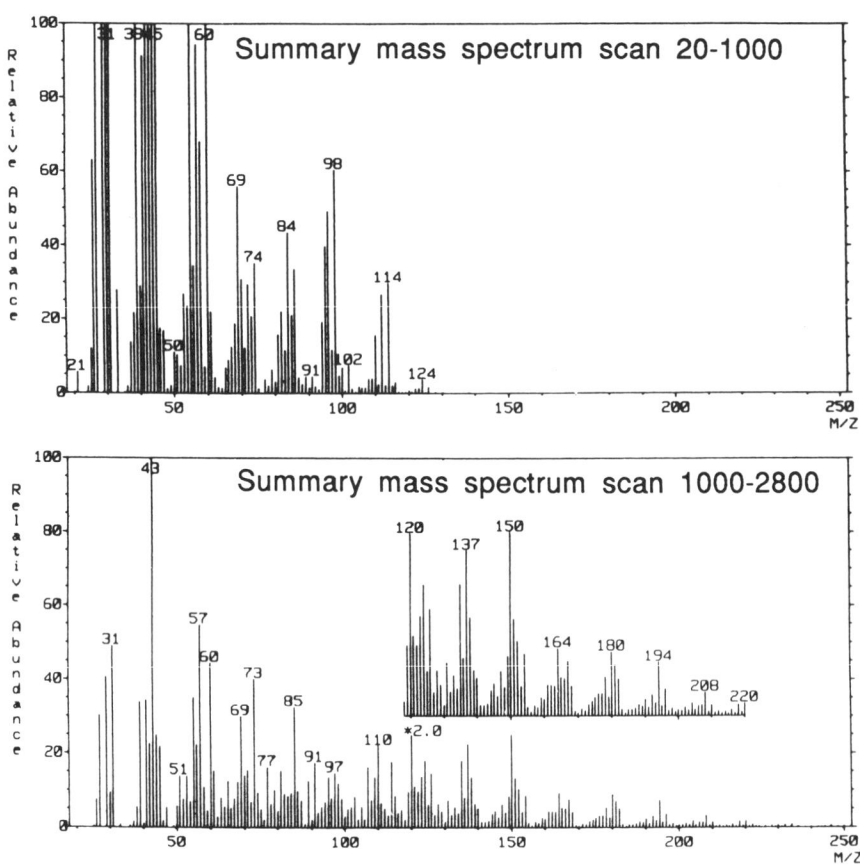

Figure 9. Time integrated spectra over selected scan ranges in OECD barley straw PYGCMS data (see Figure 7).

also in 5-hydroxymethylfurfural evolved from hexoses. The m/z 144 is from dianhydrohexoses and 1,4-dideoxy-D-*glycero*-hex-1-enopyranos-3-ulose, a specific marker compound for glucans. The m/z 73 trace shows a number of broad peaks from anhydropentoses and anhydrohexoses with 1,6-anhydropyranose (levoglucosan) as the major peak. These peaks give rise to m/z 150 and 180 in the ammonia CI spectra (Figure 6 B). The homologues of the sugars (dimers, trimers etc.) are not observed in the GCMS data because they precipitate in the glass liners around the Curie point probes (see Figure 2) and are not amenable to GC analysis. These compounds are presently subjected to HPLC analysis and off-line MS characterization.

Correlation with other data. The cell wall polysaccharides of barley straw are cellulose and xylans with low proportions of arabinofuranosyl units linked to

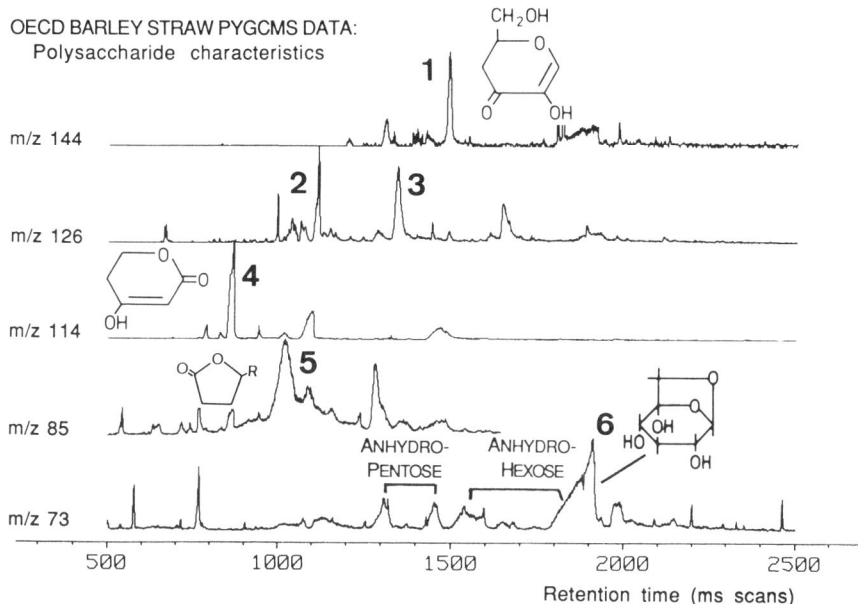

Figure 10. Mass chromatography for m/z 144, 126, 114, 85 and 73 indicative for polysaccharide marker peaks in the PYGCMS 70 eV EI data of OECD barley straw. Compounds identified are: 1 = 1,4-dideoxy-D-glycero-hex-1-enopyranos-3-ullose; 2 = several pyranone isomers (M^+ 126)[29]; 3 = 5-hydroxymethyl-2-furfural; 4 = 3-hydroxy-2-penteno-1,5-lactone; 5 = 4-hydroxymethyl-1,4-butyrolactone[32]; 6 = 1,6-anhydroglucose.

O-3 of xylose. Both *p*-coumaric and ferulic acid are ester linked to an arabinoxylan[34]. The acids are linked to the *O*-5 of the arabinofuranosyl units. Chesson *et al.*[35] could show that 70% of all alkali labile bonds in the polysaccharides are acetylated or acylated, the remaining 30% are thought to be linked to polyphenols of which nothing appears to be known. Recently truxillic acids (cyclobutyl dimers of phenolic acids) have been shown[36] to be relatively important in the cell walls as well. Evidence for cellulose and arabinoxylans is present in the PY(GC)MS data. Time resolved data acquisition may show a different behaviour of the various phenolic acids and the ether lignin. Truxillic acids will probably dissociate to vinyl phenols during pyrolysis if they are esterified to the walls. As free acids they should show up with their M^+ at 328 in EI MS data.

PYMS mapping of *Zea mays* and *Agropyron* varieties

Zea mays. Samples from early summer and late autumn specimens of three varieties of *Zea mays* i.e. Brown Midrib (BM), Eta Ipho (EI) and LG 11 (LG)

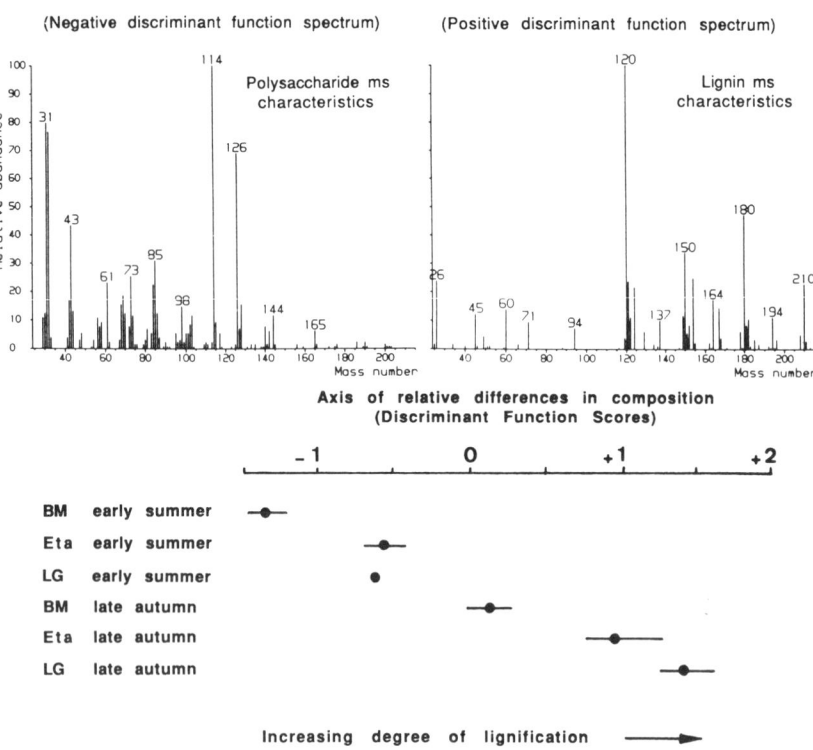

Figure 11. Discriminant function spectra and score plot for a data file of PYMS from *Ze mays* L. var. BM, Eta Ipho and LG 11 stem samples showing the relative differences i composition due to polysaccharide and lignin. PYMS: FOMautoPYMS, anal. temp. 510°C

were analysed by PYMS on the FOMautoPYMS (triplicate analyses) an compared with multivariate data analysis. The samples were a gift of Dr. F Engels from the Agricultural University in Wageningen and have been subjec of microscopic and wet chemical studies. Figure 11 shows the relativ differences between the PYMS plotted as discriminant scores (first discriminar function). A classification is obtained in which the early summer samples ar clearly separated from the late autumn samples. The variety BM separate clearly from EI and LG. The axis of relative difference in composition can b translated in polymer language by interpretation of the discriminant functio spectra for this data set. Negative scoring samples are characterized by highe intensities of the mass peaks in the spectrum with distinct polysacchari d characteristics. The spectrum correlated with positive scores shows mass peak for lignin derived phenolic compounds and m/z 60, 45 from acetic acid generate from acetyl groups. The m/z 120, 150 and 180 are indeed substantially increase in the late autumn spectra of EI and LG specimens. The m/z 120 is correlate to relatively high amounts of *p*-coumaric acid in esterified form [Emon:

unpublished results]. The distribution of the samples in the discriminant score plot is therefore interpreted as to be governed by the genetic differences between the corn varieties and by phenotypical differences as a result of an increased degree of lignification accompanied by stronger acetylation of the polysaccharides over the season.

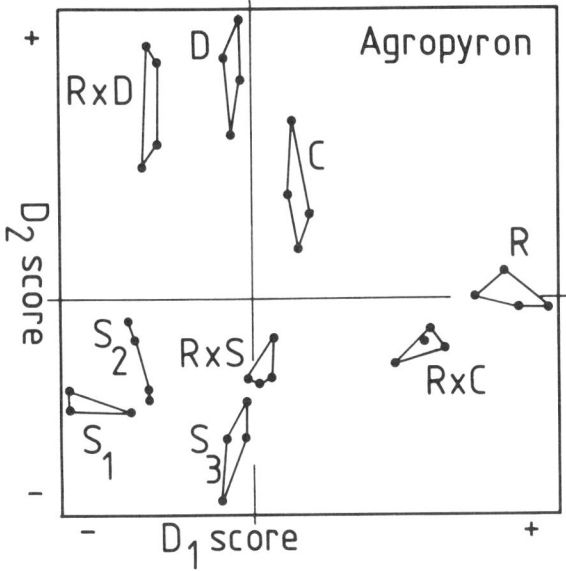

Figure 12. Map of discriminant scores of *Agropyron* species obtained by multivariate analysis of PYMS data on their second leaf. Sample codes: *A. repens* (R), *A. spicatum* (S1, S2 and S3), *A. cristatum* (C), *A. desertorum* (D) and their hydrids RxS, RxC and RxD.

Agropyron. Samples from the second leaf of *Agropyron repens* [R], *A.spicatum* [S], *A.cristatum* [C] and *A.desertorum* [D] and some of their hybrids (samples were a gift of Dr. Campbell, Utah State University) were analysed by FOMautoPYMS and multivariate analysed. Samples of similar plants have been studied earlier by Windig *et al.*[34]. Our results are shown in the form of a discriminant map (Figure 12) of the first and second discriminant function. The distribution along the main axis of discrimination (D_1) shows a differentiation of *A.spicatum, A.desertorum, A.cristatum* and *A.repens. A. spicatum* and *A.desertorum* are further differentiated by the second discriminant function (D_2). Samples from separate lots were available for *A.spicatum,* which shows the degree of biological variability (separate studies on the various leaves in several plants showed a good clustering for the first, second and third leaf, but a large difference in composition in the fourth and older leaves). The hybrids in Figure 12 cluster close to one of the parents which suggest a relatively specific distribution of genes and thus of chemical properties. Sample RxC plots close to *A.repens*, RxS plots close to *A.spicatum* and RxD plots close to *A.desertorum*.

These examples of the PYMS screening approach demonstrate the usefulness of the rapid evaluation of plant varieties. The chemical properties of existing and newly developed varieties - either in the classical way or by genetic engineering - can be profiled and compared in a semiquantitative way.

Microsampling: parenchyma and vascular bundles of *Zea mays*

Analytical pyrolysis studies are certainly useful in comparative studies of plant derived materials. The method gives preliminary information on the relative distribution of polysaccharides, lignin, cutin and other compounds, although more fundamental studies on the molecular relationship between pyrolysate composition and macromolecular structure are necessary. The method can be used for the molecular characterisation of lignin and lignified tissues. Especially the higher molecular weight pyrolysis products should give insight into the linkage of monomeric units in the polymers. The method is also valuable for characterization of cuticles[37] and suberinized tissues[10]. With this broad range in characterization potential, a combination of microscopy, histochemistry and analytical pyrolysis seems natural[36,37]. Figure 13 shows the example of PYMS of parenchyma and vascular bundles from fifth internode stem (diam. 10 mm) of mature *Zea mays* var. LG 11. The vascular bundles from the sub-epidermal sclerenchym (Figure 13 A) and from deeper inside the stem (Figure 13 B) are clearly heavily lignified but show differences in relative distribution of the lignin specific mass peaks (compare Figure 11). Especially m/z 120 from *p*-coumaric and truxillic acid groups is very intense in the highly fluorescent sub-epidermal sclerenchyma. The polysaccharide/lignin ratio appears to be different in the isolated vascular bundles deeper inside the stem in which the m/z 114 from pentosans is relatively intense. The parenchyma tissue at 1 and 4 mm in the stem (Figure 13 C and D) show the mass peaks characteristics for polysaccharides and lignin. Multivariate analysis shows that the polysaccharides in these two samples are significantly different. The parenchyma from 1 mm deep is richer in pentosan relative to hexosan, whereas this ratio is different in the parenchyma deeper in the stem (4 mm). In the latter the m/z 60, 43 and 45 from acetate is much more intense pointing to a higher degree of acetylation of the polysaccharides. The lignin pattern of the parenchyma samples is not the same either, but differs from the vascular bundles as well by lower peaks for the mixed guaiacyl-syringyl etherified lignin.

Further increases in resolution are possible by microsampling of specific tissues with a microtome. Laser desorption and surface ablation of microscopically significant areas will be possible in the future but will require new instrumental developments compared to existing instrumentation[38]. Desorption and pyrolysis of the microsample and the analysis by laser photoionization will have to be spatially separated. The confirmation of the

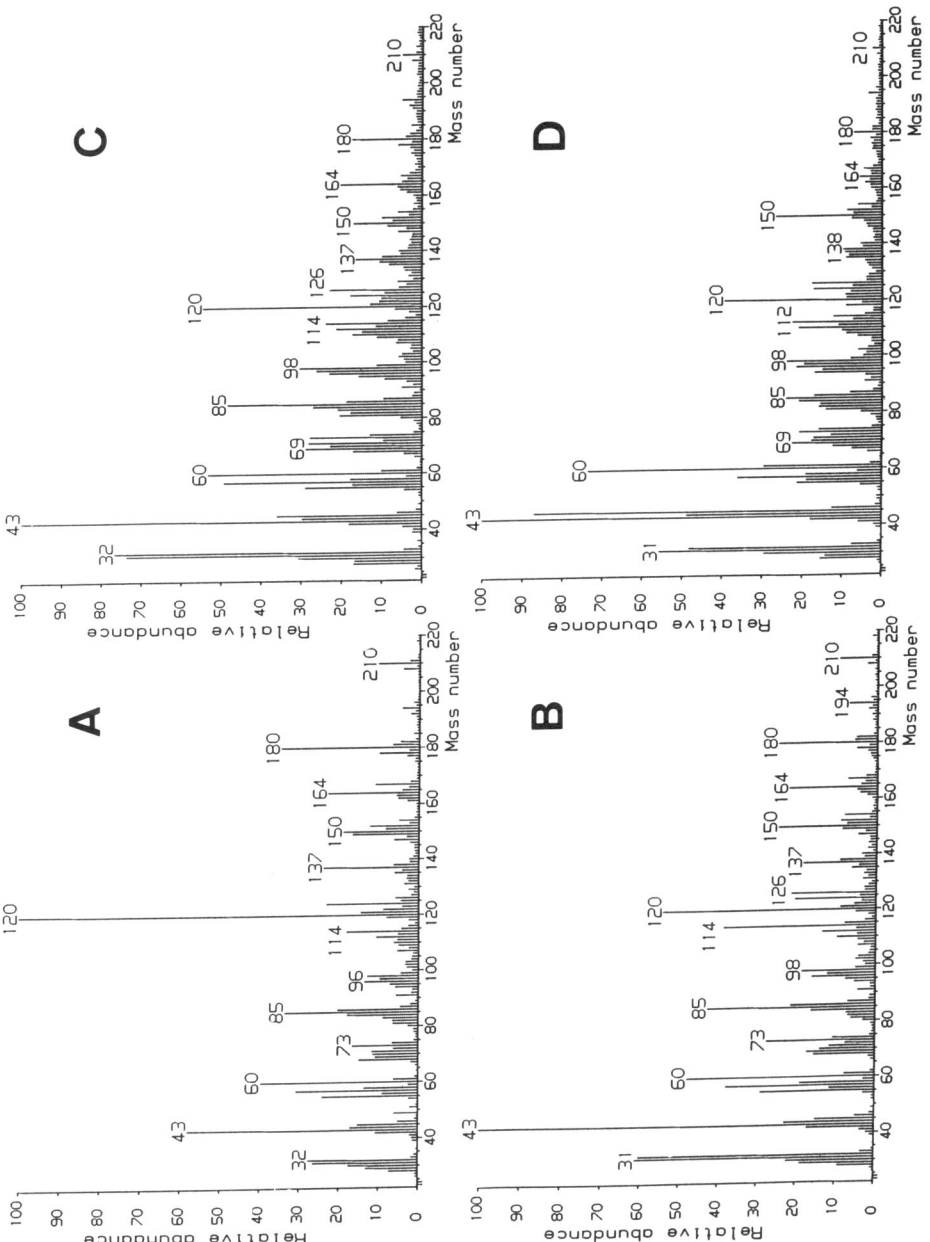

Figure 13. PYMS spectra of microsampled sub-epidermal vascular bundle (A), isolated vascular bundle in the stem (4 mm depth) (B), and parenchyma from 1 (C) and 4 mm (D) depth in the stem of mature *Zea mays* L. var. LG11 (fifth internode, 180 days old plant).

molecular structure of the desorbed molecular species will remain a special problem and may require hybrid instruments (MSMS). These new techniques should be especially useful however to answer questions on the molecular nature of the surface of plant cells.

Acknowledgements

The technical assistance of Ms. A. Tom, Mrs. B. Brandt-de Boer, Mr. G.B. Eijkel, Mr. H. van de Brink and Ir. H. Dassel is gratefully mentioned. The author is indebted to Dr. R. Hartley (AGRI, Hurley, U.K.), Dr. A. Chesson (Rowett Research Institute, Aberdeen, U.K.), Dr. W.F. Campbell and Dr. G.G. Smith (Utah State University, Logan, U.S.A.), Dr. J.van der Meer (IVVO, Lelystad N.L.), Dr. F. Engels, Dr. A. Emons and Dr. M. Willemse (Agric. Univ. Wageningen, N.L.) for samples and helpful discussions.

REFERENCES

1. Duckworth, H.E.; Barber, R.C. and Venkatasubramanian, V.S. (1986). *Mass Spectrometry,* Cambridge University Press, Cambridge.

2. Linskens, H.F. and Jackson, J.F. Eds. (1986). *Gas chromatography/Mass Spectrometry. Modern methods of plant analysis,* New series Volume 3, Springer Verlag, Berlin.

3. Maclafferty, F.W. (1983). *The interpretation of mass spectra,* University Science Books, Mill Valley, California.

4. *Eight Peak Index of Mass Spectra* (1986), Royal Society of Chemistry, Unwin Brothers, Surrey.

5. Morris, H.R., Ed (1981). *Soft Ionization Biological Mass Spectrometry,* Heyden, London.

6. Gaskill, S., Ed. (1986). *Mass Spectrometry in Biomedical Research,* Wiley and Sons, New York.

7. Meuzelaar, H.L.C.; Haverkamp, J. and Hileman, F.D. (1982). *Pyrolysis mass spectrometry of recent and fossil biomaterials,* Elsvier, Amsterdam.

8. Boon, J.J.; Pouwels, A.D. and Eijkel, G.B. (1987). Pyrolysis high resolution GCMS studies on beech wood. I. Capillary high resolution MS of a beech lignin fraction. *Transactions of the Biomedical Society* **15:** 170-174.

9. Schulten, H-R.; Simmleit, N. and Mueller, R. (1987). High temperature, high sensitivity Pyrolysis Field Ionization Mass Spectrometry. *Analytical Chemistry* **59**: 2903-2908.

10. Smeerdijk, D.G. van and Boon, J.J. (1987). Characterization of subfossil *Sphagnum* leaves, rootlets of Ericaceae and their peat by pyrolysis high resolution gas chromatography-mass spectrometry. *Journal of Analytical and Applied Pyrolysis* **11:** 377-402.

11. Pouwels, A.D.; Tom, A,; Eijkel, G.B. and Boon, J.J. (1987). Characterization of beech wood and its holocellulose and xylan fraction by pyrolysis gas chromatography mass spectrometry. *Journal of Analytical and Applied Pyrolysis* **11**: 417-436.

12. Helleur, R.J. (1987). Characterization of the saccharide composition of heteropolysaccharides by pyrolysis capillary gcms. *Journal of Analytical and Applied Pyrolysis* **11:** 297-311.

13. Saiz-Jimenez, C.; Boon, J.J.; Hedges, J.; Hessels, J.K.C. and Leeuw J.W. de (1987). Chemical characterization of recent and buried woods by analytical pyrolysis: comparison of pyrolysis data with 13C-NMR and wet chemical data. *Journal of Analytical and Applied Pyrolysis* **11:** 437-450.

14. Ishiguro, S. and Sugawara, S. (1978). Gas chromatographic analysis of cigarette smoke by the trimethylsilylation method. *Beiträge für Tabakforschung International* **9:** 218-221.

15. Radlein, D.St.A.G.; Grinshpun, A.; Piskorz, J. and Scott, D.S. (1987). on the presence of anhydro-oligosaccharides in the sirups of fast pyrolysis of cellulose. *Journal of Analytical and Applied Pyrolysis* **12:** 39-51.

16. Stout, S.A.; Boon, J.J. and Spackman, W. (1988). Molecular aspects of the peatification and early coalification of angiosperm and gymnosperm woods. *Geochimica et Cosmochimica Acta* **52:** 405-415.

17. Tromp, P.J.J.; Moulijn, J.A. and Boon, J.J. (1986). Probing the influence of K_2CO_3 and Na_2CO_3 addition on the flash pyrolysis of a lignite and a bituminous coal with Curie-point pyrolysis techniques. *Fuel* **65:** 960-967.

18. Venema, A. and Veurink, J. (1985). A new method for solvent free application of polymers and inorganic materials to ferromagnetic wires used for pyrolysis capillary gas chromatographic methods. *Journal of Analytical and Applied Pyrolysis* **7:** 207-213.

19. Meuzelaar, H.L.C.; Windig, W.; Harper, A.M.; Huff, S.M.; McClennen, W.H. and Richards, J.M. (1984). Pyrolysis mass spectrometry of complex materials. *Science* **226:** 268-274.

20. The multivariate package used at FOM is a modified and extended ARTHUR package from Infometrix, Seattle: Hoogerbrugge, R.,; Willig, S.J. and Kistemaker, P.G. (1983), *Analytical Chemistry* **55:** 1710-1712; Boon, J.J., Tom, A, Brandt, B, Eijkel, G.B. Kistemaker, P.G., Notten, F.J.W. and Mikx, F.H.M. (1984). *Analytica Chimica Acta* **163:** 193-205.

21. Brock, T.C.M.; Boon, J.J. and Paffen, B.G.P. (1985). The effects of the season and of water chemistry on the decomposition of *Nymphaea alba* L.; weight loss and pyrolysis mass spectrometry of the particulate matter. *Aquatic Botany* **22:** 197-229.

22. Bracewell, J.M. and Robertson, G.W. (1987). Characteristics of soil organic matter in temperate soils by Curie point PYMS III. Transformations occurring in surface organic horizons. *Geoderma* **40:** 333-344.

23. Hartley, R.D. and Haverkamp, J. (1984). Pyrolysis mass spectrometry of the phenolic constituents of plant cell walls. *Journal of the Science of Food and Agriculture* **35:** 14-20.

24. Valk, F.van der; Boon, J.J. and Harley, R.D. (1986). Curie-point pyrolysis-ms, -gc/ms, and -ms/ms of native, steam and ammonia treated OECD barley straw. In: J.F.J. Todd (Ed.), *Proc. of 10th international mass spectrometry conference, Advances in Mass Spectrometry,* Vol. B, p. 655-656, Wiley and Sons, New York.

25. Genuit, W. and Boon, J.J. (1985). Pyrolysis-gas chromatography-photoionization-mass spectrometry, a new approach in the analysis of macromolecular materials. *Journal of Analytical and Applied Pyrolysis* **8:** 25-40.

26. Genuit, W.A.; Boon, J.J. and Faix, O. (1987). Characterization of beech milled wood lignin by pyrolysis-gas chromatography-photoionization mass spectrometry. *Analytical Chemistry* **59:** 508-513.

27. Haider, K. and Schulten, H-R. (1985). Pyrolysis Field Ionization Mass Spectrometry of lignins, soil humic compounds and whole soil. *Journal of Analytical and Applied Pyrolysis* **8:** 317-331.

28. Coates, M.L. and Wilkins, C.L. (1987). Laser-desorption fourier transform mass spectrometry of polysaccharides. *Analytical Chemistry* **59:** 196-200.

29. Pouwels, A.D.; Eijkel, G.B. and Boon, J.J. (1988). Curie-point pyrolysis capillary gc-HR-ms of microcrystalline cellulose. *Journal of Analytical and Applied Pyrolysis.* (in press).

30. Kaaden, A. van der; Boon, J.J. and Haverkamp, J.H. (1984). The analytical pyrolysis of carbohydrates II: Differentiation of homopolyhexoses according to their linkage type by pyrolysis mass spectrometry and pyrolysis gaschromatography mass spectrometry. *Biomedical Mass Spectrometry* **11:** 486-492.

31. Faix, O.; Meier, D. and Grobe, I. (1987). Studies on isolated lignins in woody materials by pyrolysis gcms and off-line pyrolysis gc with flame ionization detection. *Journal of Analytical and Applied Pyrolysis* **11:** 403-417.

32. Aries, R.E.; Gutteridge, C.S.; Laurie, W.A.; Boon, J.J. and Eijkel, G.B. (1988). A pyrolysis mass spectrometry investigation of pectin methylation. *Analytical Chemistry* **60:** 1498-1502.

33. Ohnishi, A. and Kato, K. (1977). Thermal decomposition of tobacco cell wall polysaccharides. *Beiträge für Tabakforschung International* **9:** 147-152.

34. Mueller-Harvey, I.; Hartley, R.D.; Harris, P.J. and Curzon, E.H. (1986). Linkage of *p*-coumaroyl and feruloyl groups to the cell wall polysaccharides of barley straw. *Carbohydrate Research* **148**: 71-85.

35. Chesson, A.; Gordon, A.H. and Lomax, J.A. (1983). Substituent groups linked by alkali-labile bonds to arabinose and xylose residues of legume, grass and cereal straw cell walls and their fate during digestion in the rumen. *Journal of the Science of Food and Agriculture* **34**: 1330-1340.

36. Ford, C. and Hartley, R.D. (1988). Identification of phenols, phenolic acids, dimers and monosaccharides by gc on a capillary columns. *Journal of Chromatography* **436**: 484-490.

37. Windig, W.; Meuzelaar, H.L.C.; Haws, B.A.; Campbell, W.F. and Asay, K.H. (1983). Biochemical differences observed in pyrolysis mass spectra of range grasses with different resistance to *Labops hesperius* Uhler attack. *Journal of Analytical and Applied Pyrolysis* **5**: 183-198.

38. Nip, M.; Tegelaar, E.W., Leeuw, J.W. de; Schenck, P.A. and Holloway, P.J. (1986). A new non-saponifiable highly aliphatic and resistant biopolymer in cuticles. Evidence from pyrolysis and ^{13}C-NMR analysis of present day and fossil plants. *Naturwissenschaften* **73**: 579-585.

39. Stout, S.A.; Spackmanm, W.; Boon, J.J.; Kistemaker, P.G. and Bentley, D.F. (1988). Correlations between the microscopic and chemical changes in wood during peatification and early coalification: a canonical variate study. *International Journal of Coal Geology*. (In press).

40. Boon, J.J. and Willemse, M.T.M. (1988). Pyrolysis mass spectrometry and autofluorescence of *Zea mays* L. cell walls from parenchyma and vascular bundles. In prep.

41. Hercules, D. (1984). Solid state mass spectrometry using a laser microprobe. In: *Analytical Pyrolysis, Techniques and Applications* K. Voorhess (ed.) p.1-41. Butterworth, London.

CELL WALL AUTOFLUORESCENCE

M.T.M. WILLEMSE

Department of Plant Cytology and Morphology, Agricultural University Wageningen, Arboretumlann 4, 6703 BD Wageningen, The Netherlands

SUMMARY

The autofluorescence of cell walls or cell wall parts can be quantified by microspectroscopy. In lignified cell walls it is mainly phenol rings which give rise to autofluorescence. The intensity of autofluorescence depends on the cell wall composition and construction. The changes in autofluorescence can be used only as a coarse indicator of cell wall products as long as the construction and composition are unknown. However changes in composition such as those that occur during cell wall development or cell wall breakdown can be followed by the decrease in autofluorescence intensity or shift in wavelength. Thus when used in combination with other methods of analysis, autofluorescence can be a rapid and useful technique in cell wall research.

INTRODUCTION

Plant cell walls commonly contain products such as lignin, cutin, suberin or sporopollenin which, after being illuminated with ultraviolet light, produce primary or autofluorescence.[11] After light absorption a molecule passes into an excited state and the subsequent de-excitation may result in fluorescence, phosphorescence, photochemistry, or radiationless de-excitation. Fluorescence is the emission of photons as a result of the return of the electron to the lower orbital. The system cannot radiate more energy than it absorbed, therefore the fluorescence is at a longer wavelength than the incident absorbed light.

Molecules, which fluoresce after excitation, are characterised by the presence of double or conjugated bonds. Autofluorescence of cell walls originates mainly from the de-excitation of molecules with unsaturated bonds. However, the environment of the molecule influences the manner of de-excitation,[5,12] as demonstrated for cell walls by the presence of air or change in pH.[18] In complex structures, such as the plant cell wall, it is often difficult to relate autofluorescence to a specific molecule or even a related group of molecules. In such cases autofluorescence serves only as an indicator of the presence of a molecule or group of molecules previously identified by other chemical or histological tests. Experiments have shown that common cell wall products such as lignin, cutin, suberin or sporopollenin show autofluorescence. In some cases when one molecule is dominant in a cell wall (eg. *p*-coumaric acid), autofluorescence can be more useful, indicating both distribution and amounts present. Thus in combination with histological tests and anatomical knowledge, the use of a fluorescence microscope can provide useful information about the composition of the cell walls of different tissues. Although only a coarse indicator, autofluorescence also can be used in comparative studies to monitor changes in composition which occur during cell wall formation and breakdown, or following chemical treatment.

FLUORESCENCE SPECTROSCOPY OF CELL WALLS

An autofluorescence spectrum of a cell wall, or a part of a cell wall, can be measured with a microspectrophotometer. Depending on the sensitivity of the equipment it is possible to measure only a very small part of the cell wall. From the broad spectrum usually obtained the spectral maximum and the intensity at the spectral maximum can be selected as well as other values.[16] Generally autofluorescence occurs together with a photochemical reaction which results in a decrease, fading, or sometimes an increase of the intensity as well as a shift in wave length of the spectral maximum. Low temperature or special embedding media may suppress the fading and shift of the wavelength. It is important that spectroscopic parameters, time and all the other conditions used during the measurement should be constant if data are going to be compared.

Microspectrophotometric analysis of autofluorescence offers a value for the maximum of the wavelength, the intensity at that maximum and the changes of this intensity produced by a very small part of the cell wall. It is often difficult to relate the observed changes of the parameters to the cell wall composition and construction. Some variation in values also has to be accepted, not only because of the complexity of the cell wall, but because the fluorescence is a dynamic signal inducing photochemical reactions.

AUTOFLUORESCENCE AND LIGNIFIED CELL WALLS

In cell walls molecules with phenol rings (eg. lignin) are the main source of autofluorescence. Cell-wall formation is initiated in the cell plate after mitosis. Here vesicles fuse, a new plasma membrane is formed, and the first cell wall material, mainly containing pectins, is deposited in the first layer: the middle lamella. Subsequently criss-cross oriented, short cellulose fibrils are polymerised and laid down with hemicellulose on the plasma membrane. This primary cell wall is covered by a secondary one with parallel oriented layers of long cellulose fibrils embedded in a matrix of hemicellulose. With the formation of the secondary cell wall lignin synthesis can occur outside the plasma membrane. Irregular bound phenylpropane units are formed by oxidative polymerisation of cinnamyl alcohols catalysed by a peroxidase.[10,13] In sections of plant tissue the middle lamella of lignified cell walls generally shows a strong autofluorescence. In this layer the lignin is present as short chains of cinnamyl alcohols linked by an ether bond to the polysaccharides.[8] According to Cattesson[4] binding of lignin molecules occurs on *vic*-glycol groups. The association between the cellulose and lignin can be broken,[7] and the lignin can infiltrate from the primary wall into the secondary wall through the inter-microfibrillar spaces.[15]   Both features suggest that the random polymerisation resulting in the lignin network can form a rigid-packing and function as a physical entanglement.[8] When the inter-microfibrillar space is large, incrustation of lignin may result in a strong and compact cell wall, which is difficult to permeate, as shown by its resistance to chlor-zinc-iodine staining.[7]

In such strong and compact cell walls, autofluorescence also decreases during the lignification. The sclerenchyma of 180 day old maize stem tissues has more "lignin" than 80 day old maize stem as shown by gas chromatography[6] and pyrolitic mass spectroscopy. The intensity of autofluorescence of these 180 day old cell walls, however, is somewhat decreased. After the treatment a 1:1 mixture of 30% hydrogen peroxide and 95% acetic acid at 100°C, to dissolve the cell wall matrix, the autofluorescence intensity increased about two-fold. The process of cell wall thickening and lignification resulting in a rigid cell wall also changed the features of autofluorescence. This may be because i) the molecular condition for emission is changed; b) molecular environment of the lignin polymer does not permit the light emission and iii) the light is absorbed or partly reflected by the compact construction of the cell wall.   Such effects due to the cell wall composition and construction of secondary cell walls should be kept in mind when using autofluorescence.

CELL WALL FORMATION AND AUTOFLUORESCENCE

In comparative studies one or more spectroscopic parameters of autofluorescence can be used. The pollen wall of different species shows a characteristic autofluorescence expressed as the spectral maximum intensity. During development of the pollen wall the wavelength and intensity commonly change.   Autofluorescence parameters also can change drastically after chemical treatments of the pollen wall. Because of the complexity of the cell wall such data cannot yet be related to the changes at a molecular level.[17] In pine pollen the main fluorescent component seems to be p-coumaric acid.[14]

The different stem cell walls and wall layers of *Populus* show slight differences in the spectral maximum and intensity of autofluorescence. Samples of wood, show their own autofluorescence characteristics differing in their spectral maximum and rate of fading even after chemical treatments.[18]

These experiments demonstrate no direct relationship between the cell wall composition and construction and autofluorescence. The spectral maximum produced is only a coarse indicator of the molecular differences. In general, sporopollenin and cutin commonly show a more yellow fluorescence, while suberin and lignin a more blue one. The intensity can be a measure of the total number of excited molecules present in a fixed cell wall area, but it is difficult to use the method quantitatively when the construction and composition of the cell wall is unknown. In parenchyma cell walls of growing maize, autofluorescence is produced primarily by ferulic and p-coumaric acids. The increase in autofluorescence intensity during the first 70 days is a measure of the synthesis of both ferulic and *p*-coumaric-acids during the development of this wall.[19]

The fading, positive or negative, is a result of photochemical reactions. The velocity of this reaction can be illustrated by quantification of the fading, but the nature of the photochemical reactions that occur remain unknown.

Present quantitative data from autofluorescence spectra indicate some characteristics of cell walls or cell wall formation, but the value of the data produced is still limited. However when combined with an increasing knowledge of cell wall composition and construction, autofluorescence is likely to make an increasing contribution to cell wall analysis.

CELL WALL BREAKDOWN AND AUTOFLUORESCENCE

In cell wall digestion, autofluorescence can be used as a marker of the changes occurring in the phenolic compounds present in the cell wall.

An example of the use of autofluorescence to study cell wall breakdown is shown in Figure 1. Both data from the fungal degradation of coconut fibre and the dioxane: 0.1% HCl (9:1 v/v) extraction of the secondary xylem of *Picea abies* demonstrate the changes occurring in the lignin faction of the wall.[18]

The fungus *Trametes* sp. produces a lignolytic enzyme affecting the coconut fibre. The fibre looses about half of its tensile strength after treatment.[2] The autofluorescence shows a decrease in intensity, suggesting a disappearance of the lignin component. As the stability of the spectral maximum suggests that any molecular change is absent, autofluorescence can be used to follow fibre breakdown and, indirectly, the activity of the lignolytic enzymes.

Removal of lignin from the secondary xylem of the *Picea abies* tracheids has a more drastic effect. After 25 minutes of boiling in dioxane-water, a part of the lignin is removed, which is indicated by the now negative phloroglucinol reaction. Before heating, the treatment showed a slight shift in spectral maximum. During boiling, the spectral maximum shifted to yellow during the first 15 minutes and thereafter to blue. Such a feature can be an indicative of the loss of a distinct group of molecules. The intensity also decreased confirming the extraction of phenolic material. The experiment illustrates the velocity of chemical breakdown of the lignin, but the explanation for the changing spectrum remains difficult.

In permanganate-delignified sections of Bermuda grass stems, the autofluorescence intensity decreased to zero in the parenchyma. In the sclerenchyma the decrease was not so drastic, probably due to a lesser penetration of the permanganate. This sclerenchyma is also less susceptible to digestion.[1]

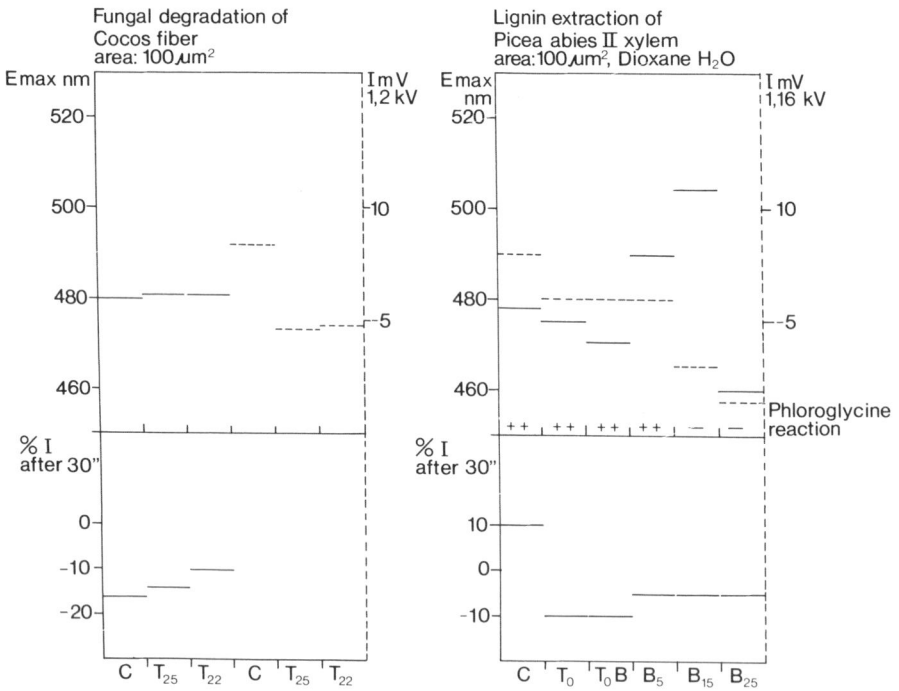

Figure 1. Fungal degradation of coconut (cocos) fibre. T = *Trametes* sp. Stock 22 and 25; C = control untreated fibre; Emax nm = maximal wave length; Imv = intensity in mV at Emax measured at 1.2 kV; %I - fading after 30 sec in percentage. Lignin extraction of *Picea abies* secondary xylem. C = control; T_0B = 1 second in boiling mixture; $B_{5,15,25}$ = after 5, 15, 25 min boiling. Only mean values are noted.

Autofluorescence was also measured on untreated and ammonia treated straw sections. Sections were treated according to the NDF and ADF method or were delignified with potassium permanganate.[3] Table 1 shows the results obtained for the sclerenchyma and parenchyma cell walls.

In the delignified cell walls the spectral maximum shifted to a lower wavelength in the untreated sclerenchyma and parenchyma to about 450 nm and the intensity decreased. These results indicated that lignin had been extracted leaving a residue with about 5-10% of the original intensity. A residue of autofluorescent molecules clearly persisted after delignification. The increase in intensity after neutral detergent extraction was probably due to the modification of the compact sclerenchyma wall, which changed the molecular environment allowing them to fluoresce. After ammonia treatment the intensity of autofluorescence in the sclerenchyma was lower, indicating that the wall probably had lost molecules capable of autofluorescence. In contrast, the parenchyma wall did not show the same lowered intensity after ammonia

TABLE 1. Intensity of autofluorescence (I_o, mV) and wavelength of maximum emission (E_{max}, nm) of sclerenchyma (S) and parenchyma (P) cell walls in 100 μm sections of untreated (U) and ammonia-treated (A) barley straw before and after chemical treatment

Straw sample	Fraction	S		P	
		E_{max}(nm)	I_o(mv)	E_{max}(nm)	I_o(mv)
U	None (control)	472	3.1 ± 0.1	455	2.9 ± 0.2
	NDF	470	5.0 ± 1.0	458	6.0 ± 2.0
	ADF	465	2.7 ± 0.2	453	4.2 ± 1.0
	Delignified	450	0.5 ± 0.1	450	0.3 ± 0.2
A	None (control)	475	1.3 ± 0.1	478	2.7 ± 0.3
	NDP	490	3.0 ± 0.1	485	5.5 ± 0.4
	ADF	463	1.0 ± 0.2	463	1.8 ± 0.5
	Delignified	450	0.4 ± 0.1	445	0.2 ± 0.1

treatment. The differences in wavelength may be the result of changes in the environment such as the pH, or to the molecules themselves. Autofluorescence intensity indicates the influence of the extractive methods applied to the cell wall and can be compared with the digestibility of the material.

CONCLUSION

Autofluorescence is easy to detect and can be applied to the analysis of small segments of a cell wall. The method is simple, rapid and needs only a very small sample. When used in combination with other methods of cell wall analysis autofluorescence can provide a useful technique for monitoring change. Thus autofluorescence is potentially a useful marker for changes which occur during cell degradation, following chemical treatments or during cell-wall formation. However, the interpretation of spectroscopic data remains difficult. Secondary fluorescence is an open field, yet to be applied to plant cell wall research.

REFERENCES

1. Akin, D.E.; Willemse, M.T.M. and Barton, F.E. II (1985). Histochemical reactions, autofluorescence, and rumen microbial degradation of tissues in untreated and delignified bermudagrass stems. *Crop Science* **25**: 901-905.

2. Antheunissen, J. (1980). The breakdown of coconut fibers by *Trametes* sp. and *Stereum rugosum*. *Journal of Applied Microbiology* **26**: 167-170.

3. Barton, F.E. II and Akin, D.E. (1977). Digestibility of delignified forage cell walls. *Journal of Agricultural and Food Chemistry* **25**: 1299-1303.

4. Catteson, A.M. (1983). A cytochemical investigation of the lateral walls of *Dianthus* vessels differentiation and pit-membrane formation. IAWA Bulletin **4:** 89-101.

5. Clayton, R.K. (1977). *Light and Living Matter.* Vol. 1. *The Physical Part.* R.E. Krieger Publ. Comp. Huntington, New York. 148 p.

6. Emons, A.C. and Engels, F.M. (1987). Phenolic acids of maturing stems of maize cultivars differing in digestibility. *Acta Bot. Neerl.* **36:** 23.

7. Frey-Wyssling, A. (1959). *Die pflanzliche Zellwand.* Springer Verlag, Berlin, Göttingen, Heidelberg, 367 p.

8. Fry, S.C. (1986). Cross-linking of matrix polymers in the growing cell walls of Angiosperms. *Annual Review of Plant Physiology.* **37:** 165-186.

9. Goering, H.K. And van Soest, P.J. (1970). *Forage fiber analyses* (apparatus, reagents, procedures, and some applications). Agr. Handbook N°. 379, USDA, Washington D.C.

10. Goldberg, R.; Catteson, A.M.; Czaninski, Y. (1981). Histochemical and biochemical characteristics of peroxydases involved in lignification processes in poplar. In *Cell Walls* 81. D.G. Robinson and H. Quader (eds.). Wissenschaftliche Verlagsgesellschaft, Stuttgart pp. 251-260.

11. Goodwin, R.H. (1953). Fluorescent substances in plants. *Annual Review of Plant Physiology* **4:** 283-304.

12. Lakowicz, J.R. (1983). *Principles of fluorescence microscopy.* Plenum Press, New York, London. 496 p.

13. Northcote (1985). Control of cell wall formation during growth. In *Biochemistry of Plant Cell Walls.* C.T. Brett and J.R. Hillman (eds.) Cambridge Univ. Press Cambridge, pp. 177-199.

14. Schulze Osthoff, K. And Wiermann, R. (1987). Phenols as integrated compounds of sporopollenin from *Pinus* pollen. *Journal Plant Physiology* **131:** 5-15.

15. Siegel, S.M. (1962). *The Plant Cell Wall.* International Series of Monographs on Pure and Applied Biology. Pergamon Press, Oxford.

16. Van Gijzel, P. (1974). Topics in UV-fluorescence microspectrophotometer. *II: Fresh and fossil plant substances.* University of Nijmegen, The Netherlands.

17. Willemse, M.T.M. (1972). Changes in the autofluorescence of the pollen wall during microsporogenesis and chemical treatments. *Acta Botanica Neerlandica* **21:** 1-16.

18. Willemse, M.T.M. (1981). Changes in autofluorescence of lignin. In *Cell Walls 81.* D.G. Robinson and H. Quader (eds.) Wissenschaftliche Verlagsgesselshaft, Stuttgart pp.242-250.

19 Willemse, M.T.M. And den Outer, R.W. (1988). Stem anatomy and cell wall autofluorescence during growth of three maize (*Zea mays* L.) cultivars. *Acta Botanical Neerlandica* **37:** 39-47.

LIGHT MICROSCOPY AND HISTOLOGY OF LIGNOCELLULOSE RELATED TO BIODEGRADATION

D.E. AKIN

R.B. Russell Agricultural Research Center, Agricultural Research Service, United States Department of Agriculture, P.O. Box 5677, Athens, Georgia 30613, U.S.A.

SUMMARY

Histological stains for lignin indicate the site, and to some extent, variations of the lignin component in plant cell walls. Further, various stains can indicate an assessment of the biodegradability of specific plant tissues. Acid phloroglucinol reacts with the most resistant tissues to biodegradation (e.g. xylem cells), while chlorine-sulphite stains the lignocellulosic fibre less resistant to breakdown (e.g., leaf blade sclerenchyma and stem parenchyma). Diazonium salts react with a variety of phenolic acids and aldehydes, and may be useful to differentiate various compounds. Research is needed to specify the organization (e.g., chemical bonding) and composition of microsites within the plant cell walls that characterize a particular response to biodegradation.

INTRODUCTION

Histological stains for lignin have been used for several years. A listing of the more widely used ones and their functional groups for reaction is shown in Table 1. Acid phloroglucinol and chlorine-sulphite are two stains that have been useful in identifying tissues resistant to degradation by rumen microorganisms[3]. The procedures for these stains can be undertaken fairly rapidly, but chlorine-sulphide can give false negative results with samples stored for extended times or if chemicals are not fresh. Azure B, listed as a possible lignin stain[11] has not been widely used. The procedure for azure B takes several hours, and the colour will fade upon storage. Diazonium salts have been used to locate lignin in leaves and stems of annual ryegrass[9], but this potentially useful stain has not been widely applied.

Much of the information on lignin stains appears to have come from research on woody plants[12]. Forages are lower in lignin concentration compared with woody plants, and the phenolic compounds and organization within the cell wall are still not well understood in grasses[10]. Therefore, extrapolation of results on woody lignocelluloses to forages may not be appropriate in all cases. In view of this idea, the mechanisms involved in lignin stain reactions should be considered. Acid phloroglucinol is reported to react with coniferaldehyde groups in lignin giving a deep purple colour[12]. Clifford[8] reported that various formulations for acid phloroglucinol all detected most aldehydes, with various colour responses for different aldehydes and those of cinnamaldehyde compounds giving a purple colour. Sarkanen and Ludwig[12] reported that acid phloroglucinol "had universal application to all lignins, although the reaction may be weak or even absent in lignins containing high amounts of syringyl propane units".

TABLE 1. Histological Stains for Lignin

Reagent	Functional Group
Acid Phloroglucinol	Coniferaldehyde
Chlorine-Sulphite	Syringylpropane
Maule Reagent	Syringlypropane
Azure B	Phenolic hydroxyl
Diazonium Salts	Phenolic compounds

Chlorine-sulphite and the Maule reaction both are reported to indicate lignin in which large amounts of syringyl propane units are present[12]. Stafford[14], using histological and biological techniques, found different types of lignin in various tissue types, and she reported that chlorine-sulphite lignin was particularly prevalent in leaf blade sclerenchyma. The mechanism for the chlorine-sulphite reaction is not known[12].

STAINING REACTIONS OF PLANT CELL WALLS

In order to evaluate the stains for specific constituents, an ordered series of cinnamic acids, benzoic aldehydes and benzoic acids and other related phenolic compounds were tested with the histological stains for lignin. Reactions to phenolic compounds, which have been found in plant cell walls, are shown in Table 2. As can be seen from the abbreviated list, acid phloroglucinol and chlorine-sulphite do give various reactions with different compounds. Theoretically, the potential may exist for differentiation of some compounds with stains, but the reaction within the cell wall must be more extensively evaluated. Azure B does not react with any of the pure compounds or lignin complexes from the cell wall, but this stain did show a blue-green colour for lignin isolated from bermudagrass (Table 2) and from wheat straw and brown-midrid sorghum (not shown). In contrast, the lignins from tall fescue and normal sorghum gave a green colour and no reaction, respectively. The diazotized sulphanilic acid (and other diazonium salts not shown) reacted with pure phenolic compounds and all lignin fractions, giving different reactions for some compounds. It would perhaps be difficult to differentiate the various degrees of orange and red in the light microscope to identify specific compounds, but the diazonium salts might be more sensitive to low molecular weight phenolic compounds (i.e., ferulic and *p*-coumaric acids) in graminaceous plants[1].

The application of these lignin stains to forage tissue is shown for Bermudagrass (*Cynodon dactylon*) stems in Table 3. The two tissues shown give reactions representative for all tissues in these stems. The upper (second from the apex) and lower (fifth from the apex) internodes differ in their reactions, with the lower internode showing greater lignification. No definite reaction for lignin occurred in the upper parenchyma, although a slight reaction occurred with diazotized sulphanilic acid, but the sclerenchyma ring tissue shows a slight to definite reaction with all three stains. In contrast, all tissues were lignified in the lower internode, with parenchyma giving a reaction with chlorine-sulphite and diazotized sulphanilic acid and a stronger reaction in the centripetal cells; sclerenchyma was positive for lignin only with acid phloroglucinol.

TABLE 2. Reaction of Phenolic Compounds to Histological Stains

	Histological Stain			
Compound	Acid Phloroglucinol	Chlorine Sulphite	Azure B	Diazotized Sulphanilic Acid
Syringaldehyde	Red	0	0	Orange
Ferulic Acid	0	Pink	0	Red
Vanillin	Yellow (weak)	Yellow	0	Orange-Red
p-Coumaric Acid	0	0	0	Red-Orange
Methyl p-Coumarate	0	0	0	Red-Orange
Bermudagrass lignin	Yellow	Yellow	Blue-Green	Brown-Orange
Lignin-Carbohydrate complex	Orange	Orange	0	Orange-Brown

The relationship of parenchyma and sclerenchyma to biodegradation is shown in Plate 1. The sclerenchyma is totally resistant to degradation in both upper and lower internodes, with the parenchyma totally degraded in the upper but only partially degraded in the lower internodes.

Other data on upper (immature) and lower (mature) internodes in mature Bermudagrass plants provided an assessment of chemical and quality variation (Table 4; Ref. 6). While the percentage of tissue types did not change with internode position, the amount of lignin increased and the biodegradation decreased in the lower internodes. These data verify and expand the information in Table 3 and Plate 1. The greatest change in a structural sense related to quality then, is that the parenchyma increased in chlorine-sulphite lignification with maturity.

These data as well as those on other plants[3] have provided information on a consistent trend in histochemical reactions and tissue biodegradation. The most resistant tissues recognized in forage grasses are those that stain with acid phloroglucinol. The vascular xylem is a good example of this, where no

TABLE 3. Histological Reactions for Lignin in Bermudagrass (*Cynodon*) Stem Tissue

Position	Tissue	Reaction in tissue with:		
		Acid Phloroglucinol	Chlorine Sulphite	Diazotized Sulphanilic Acid
Upper	Sclerenchyma	±	+	+
	Parenchyma	-	-	±
Lower	Sclerenchyma	+	-	-
	Parenchyma	-	+	+

[a]+ = definite; ± = slight; - = no reaction.

degradation occurs even after alkali treatment[13]. In contrast, in the leaf blade sclerenchyma a definite chlorine-sulphite reaction occurs, but partial biodegradation can occur with rumen bacteria[4], and a substantial amount is degraded with rumen fungi[5]. The chlorine-sulphite reaction also occurs in tissues that become lignified with maturity, including the stem parenchyma. In warm-season grasses, living tissues that become less degradable during environmental stress often give a chlorine-sulphite positive test[2].

TABLE 4 Fibre Characteristics of Bermudagrass (*Cynodon*) Stems

Position	Tissue Type		Acid Detergent	Permanganate	
	Sclerenchyma (%)	Parenchyma (%)	Lignin (%)	Lignin (%)	IVDMD (%)
Upper	56	39	34	5	61
Lower	60	37	44	8	45

From Akin *et al.*[6]

Studies on the autofluorescence of parenchyma and sclerenchyma cells in mature Bermudagrass stems indicated a higher intensity in the sclerenchyma[7]. Upon treatment with permanganate, the parenchyma tissue was completely delignified and biodegraded, while the autofluorescence of sclerenchyma was reduced 27% and these cell walls still resisted degradation.

It appears that the two lignin stains are indicative of lignocelluloses that vary in their resistance to biodegradation. However, since identification of components

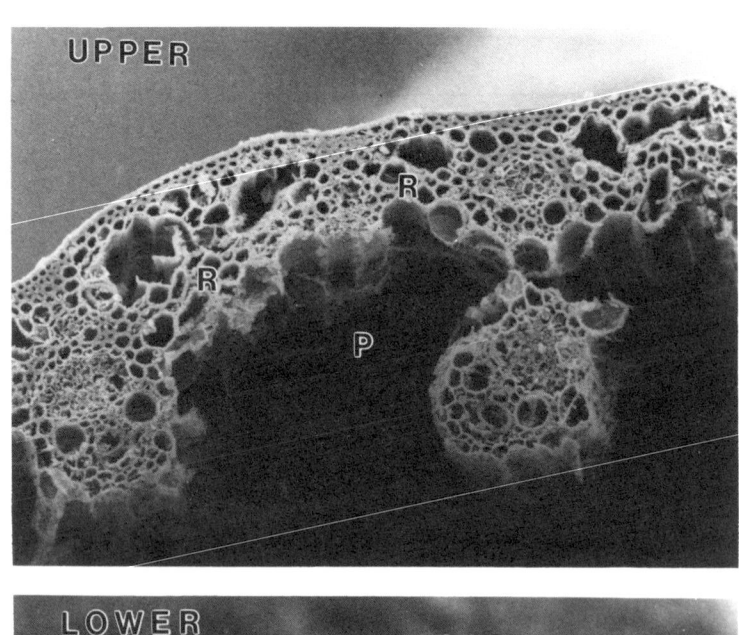

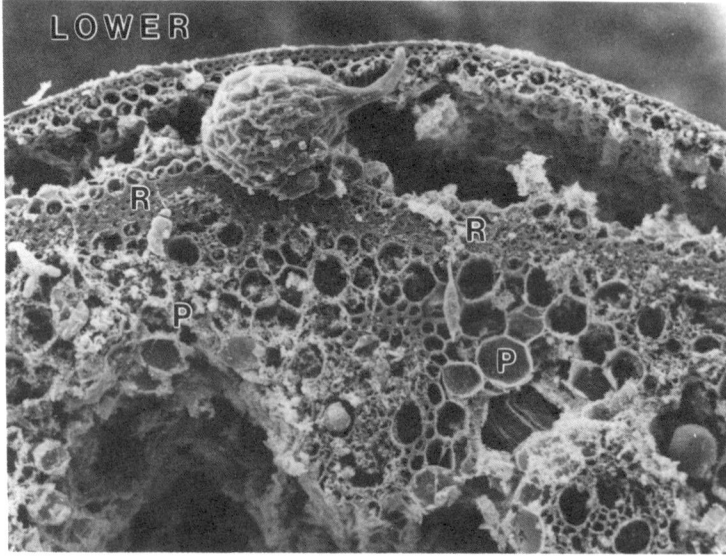

Plate 1. Upper and lower Bermudagrass stems after incubation with rumen microorganis for 96 hours. R = sclerenchyma ring. P = parenchyma.

within cell walls is difficult, it is not clear if the lignin stains indicate differences in composition of lignin or in organization and binding of phenolic components within the cell wall.

The following conclusions are made:

1. Lignin stains indicate gross sites and types of lignocellulose.

2. Diazonium salts have known reactions for phenolics and may indicate lower amounts or smaller molecular weight compounds than other stains.

3. Techniques are needed to identify and quantitate phenolic types in specific plant cell walls.

REFERENCES

1. Akin, D.E. (1986). Interaction of ruminal bacteria and fungi with southern forages. *Journal of Animal Science* **63**: 962-977.

2. Akin, D.E.; Barton, F.E. II and Colman, S.W. (1983). Structural factors affecting leaf degradation of old world bluestem and weeping love grass. *Journal of Animal Science* **56**: 1434-1446.

3. Akin, D.E. and Burdick, D. (1981). Relationships of different histochemical types of lignified cell walls to forage digestibility. *Crop Science* **21**: 577-581.

4. Akin, D.E.; Burdick, D. and Michaels, G.E. (1974). Rumen bacterial interrelationships with plant tissue during degradation revealed by transmission electron microscopy. *Applied Microbiology* **27**: 1149-1156.

5. Akin, D.E. and Rigsby, L.L. (1987). Mixed fungal populations and lignocellulosic tissue degradation in the bovine rumen. *Applied and Environmental Microbiology* **53**: 1987-1995.

6. Akin, D.E.; Robinson, E.L.; Barton, F.E. II and Himmelsbach, D.S. (1977). Changes with maturity in anatomy, histochemistry, chemistry and tissue digestibility of bermudagrass plant parts. *Journal of Agricultural and Food Chemistry* **25**: 179-186.

7. Akin, D.E.; Willemse, M.T.M. and Barton, F.E. II (1985). Histochemical reactions, autofluorescence, and rumen microbial degradation of tissues in untreated and delignified bermudagrass stems. *Crop Science* **25**: 901-905.

8. Clifford, M.N. (1974). Specificity of acidic phloroglucinol reagents. *Journal of Chromatography* **94**: 321-324.

9. Harris, P.J.; Hartley, R.D. and Barton, G.E. (1982). Evaluation of stabilized diazonium salts for the detection of phenolic constituents of plant cell walls. *Journal of the Science of Food and Agriculture* **33**: 516-520.

10. Hartley, R.D.; Whatley, F.R. and Harris, P.J. (1988). 4,4'-dihydroxytruxillic acid as a component of cell walls of *Lolium multiflorum*. *Phytochemistry* **27**: 349-351.

11. Jensen, W.A. (1962). *Botanical Histochemistry.* W.H. Freeman and Company, San Francisco.

12. Sarkanen, K. V. and Ludwig, C.H. (1971). Definition and nomenclature, p. 1-18. In K.V. Sarkanen and C.H. Ludwig (ed.) *Lignins: occurrence formation, structure and reactions.* Wiley-Interscience, New York.

13. Spencer, R.R. and Akin, D.E. (1980). Rumen microbial degradation of potassium hydroxide-treated coastal bermudagrass leaf blades examined by electron microscopy. *Journal of Animal Science* **51**: 1189-1196.

14. Stafford, H.A. (1962). Histochemical and biochemical differences between lignin-like materials in *Phleum pratense L. Plant Physiology* **37**: 643-649.

ELECTRON MICROSCOPY AS A METHOD TO EVALUATE STRUCTURE AND DEGRADATION OF PLANT CELL WALLS

E. GRENET

Unité de l'Ingestion, Station de Recherches sur la Nutrition des Herbivores, INRA, Centre de Recherches de Clermont-Ferrand, Theix, 63122 Ceyrat, France

SUMMARY

The structure and degradation of plant cell walls can be evaluated by chemical methods, but such methods measure only the contents of cell wall components. Physical methods, such as electron microscopy, provide information on the distribution of such constituents both within and between cell walls. The ease of breakdown of different plant tissues and their selective attack by microorganisms in the rumen have been visualised by electron microscopy, which provides an understanding of the differences in digestibility that have been observed. The relationship between rumen microorganisms and plant tissues, and in particular, the adhesion of bacteria to cell walls have been demonstrated by this technique, which also made possible the relatively recent discovery of anaerobic fungi in the rumen. The limitations of electron microscopy reside in the necessity to select appropriate samples that are representative of the plant studied, and the present impossibility of obtaining quantitative results. Its future development is likely to be linked to that of immunocytochemistry which should permit the study of cell walls constituents and microbial enzyme secretion in far greater detail than is possible at present.

INTRODUCTION

The microorganisms present in the rumen and the caecum of ruminants make it possible for these animals to feed on lignocellulosic substrates. This food can only leave the rumen when broken into small particles (1 mm in the case of sheep). Reduction in particle size is achieved by a combination of mastication and microbial degradation. Food is thus retained in the rumen for time periods which increase with difficulty in degradation. The intake of feeds which remain for a long time in the rumen is reduced, as the presence in the stomach of undigested residues restricts further feeding by the animal. Chemical analysis of feeds cannot, alone, explain all the aspects of digestion by ruminants. It provides information about cellulose, hemicelluloses, pectin and lignin contents but it does not provide any knowledge about the distribution of these constituents in the cell walls. For these reasons modern physical methods such as light and electron microscopy are of particular interest. They make it possible to show the presence of different plant tissues having cell walls with different compositions. Only a few research teams based in the U.S.A., Canada, Australia, Holland, U.K. and France, so far use such methods.

The structure of plant cell walls and their breakdown can be studied by light microscopy (LM), scanning electron microscopy (SEM) and transmission electron microscopy (TEM). The three techniques each provide complementary information. SEM gives a resolution of 5 to 50 nm, which being much greater than LM (200 nm) permits magnifications of up to x 100,000 compared with x 2000 for LM. In addition it gives a greater depth of field. TEM, with a resolution of below 1 nm gives enlargements of x 200,000 or more. LM allows rapid studies of the different tissues and their respec;tive proportions, the size

of the cell walls and the determination of the presence or absence of lignin. The great depth of field of SEM makes it possible to determine tissue losses on the surface of samples following degradation. It provides information on the relationship between microorganisms, particularly fungi, and plant cell wall. TEM provides unique observations on both the structure of plant cell walls and the mode of degradation by rumen microorganisms.

STUDY OF CELL WALL STRUCTURE

For SEM observations, plant samples - stem sections 0.5 cm in length for a given internode; strips of leaf 0.5 cm wide; rings cut in seed teguments of oil or protein crop seeds - are fixed in glutaraldehyde, progressively dehydrated in alcohol then dried to the critical point before being metallised. All surfaces of the sample, except that adhering to the specimen stub can be examined. The technique provides a good idea of spacial configuration.

SEM observation of a transverse maize stem section shows the disposition of different tissues[1]: epidermis, sclerenchyma, xylem and phloem and, in addition, the relative thickness of the walls of sclerenchyma and bundle sheath fibres compared to those of parenchyma (Plate 1). Observation of the stem of the mutant "brown midrib" shows that the walls of the sclerenchyma are thinner than those of normal maize. LM examination of tissues after specific staining of lignin (acid phloroglucinol stains lignified tissues rich in cinnamaldehyde groups such as xylem and Maüle's reactant stains walls rich in syringyl groups such as sclerenchyma) provides complementary information.

SEM observation of soya bean seed tegument[2], provides spectacular images since it is possible to represent the three dimensional structure of this tegument, impossible by any other method. The tegument is formed of several layers of cells; the most external being the palisade layer, followed by the column cell layer, each cell of which resembles a column, then the parenchyma and the aleurone layers. The column cells are few in number and widely separated by empty spaces (Plate 2).

Finally, LM or SEM observations show that the leaf blades of warm-season grasses which are less digestible than those of cool-season grasses, contain a higher proportion of tissues, such as epidermis and parenchyma bundle sheath, which can be degraded only slowly compared with those tissues broken down rapidly, such as phloem and mesophyll[3].

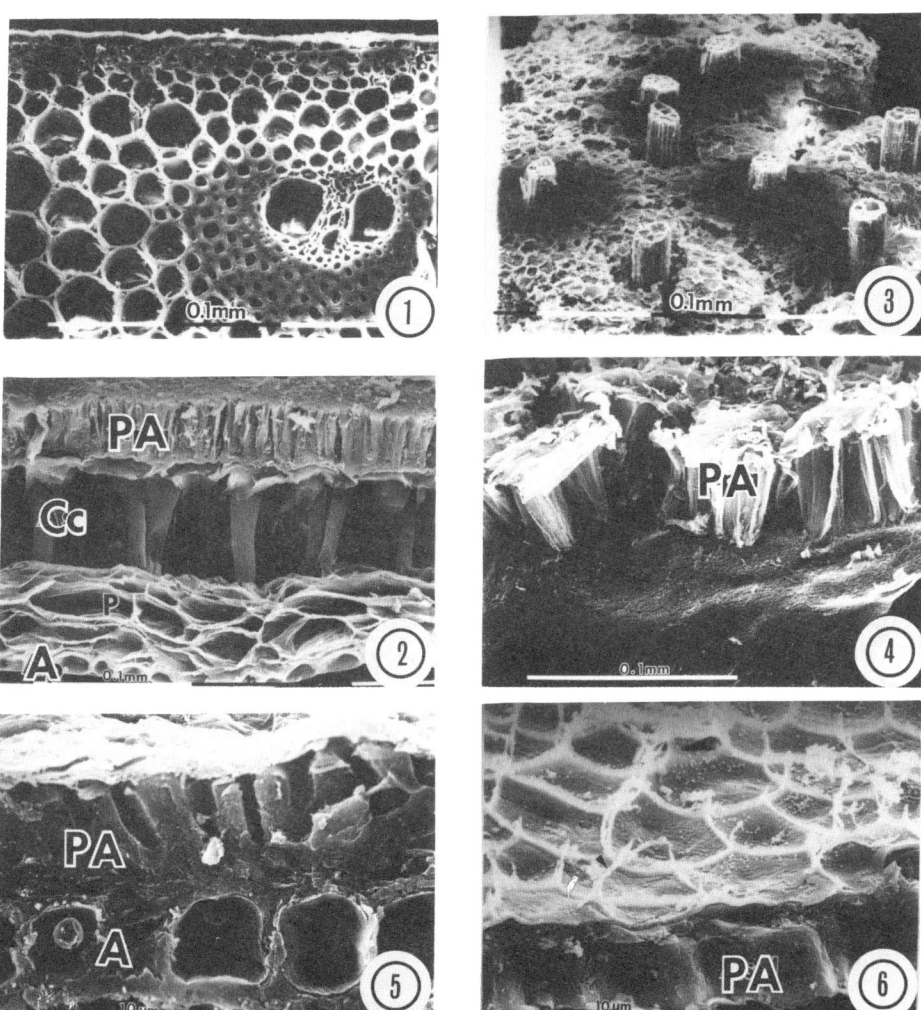

Plate 1. Transverse section of maize stem showing the epidermis, the sclerenchyma and the parenchyma with a vascular bundle. **Plate 2.** Transverse section of soya bean seed tegument showing the palisade layer (PA), the column cells (Cc), the parenchyma (P) and the aleurone layer (A). **Plate 3.** Maize stem after 48 h incubation in the rumen. The parenchyma has been degraded but the vascular bundles remain intact. **Plate 4.** Soya bean seed tegument after incubation in the rumen for 72 h. The cell layers have been degraded except some cells of the palisade layer (PA) and the aleurone layer (A). **Plate 5.** Transverse section of soya seed tegument showing the palisade layer (PA) and the aleurone layer (A). **Plate 6.** Rape seed tegument after incubation in the rumen for 72 h. The palisade layer (PA) remains intact with a part of the walls of the aleurone layer attached to it (arrow).

A combination of SEM observations and energy dispersive X-ray analysis (EDXA) have shown the localisation of silica in grass cell walls[4]. It is found in the adaxial epidermal cells, in particular above vascular bundles and in abaxial epidermal cells. Its presence inhibits rupture of the cuticule, which is necessary for microbial degradation of the epidermis[5]. Combination of EDXA with SEM or TEM offers also a semi-quantitative observation on the distribution of phenolics in cell walls[6].

For TEM studies, embedding samples in resin makes it possible to obtain ultra-thin sections of 70 to 80 nm. Progress in the ultrastructural cytochemistry of polysaccharides and lignin have made possible *in situ* detection of some plant cell wall components[7]. Detection of polysaccharides with *vic*-glycol groups can be obtained using the PATAg reaction[8], acid polysaccharides by LUFT's technique[9] with ruthenium red, and methylated acidic polysaccharides can be detected using the ferric hydroxylamine reaction[10]. Lignin can be detected either by fixation in potassium permanganate or by fixation in ethanol followed by the reaction of Coppick and Fowler[11]. These methods have been used by Engles and Brice[12] to demonstrate the presence of a tertiary wall in barley straw, covered by a warty layer, which is particularly resistant to breakdown.

STUDY OF PLANT CELL WALL DEGRADATION

SEM and TEM both provide considerable information about microbial degradation of plant cell walls which can complement measurements made of the rates of disappearance of dry matter and of cell walls.

Thus it has been possible to follow the progressive degradation of maize stems by SEM. Samples 0,5 cm length, were enclosed in nylon bags and placed in the rumen for 8, 24, 48 and 72 hours. Phloem was the first to disappear followed by the internal parenchyma of the stem and then the external parenchyma. The bundle fibres, xylem and sclerenchyma, resisted degradation[1]. The great depth of field of the SEM makes it possible to observe bundles which emerge from the surface of the sample after 48 hours, following the progressive degradation of the surrounding parenchyma (Plate 3). Comparison of the mutant brown midrib with the normal showed that the parenchyma of the mutant was degraded more rapidly than that of the normal maize.

The same technique was used to observe the degradation of grass leaf blades, which occurred in the order mesophyll, phloem > epidermis, parenchyma bundle sheath > sclerenchyma > lignified vascular tissues[13]. Observations with TEM complement those made by SEM. Thus it could be shown that only the surface of sclerenchyma cell walls are degraded by microorganisms which are attached to them[14]. Occasionally filamentous bacteria may give more intense degradation[15].

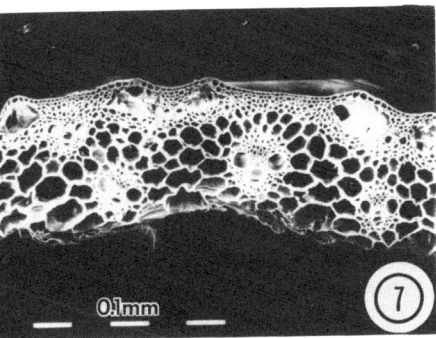
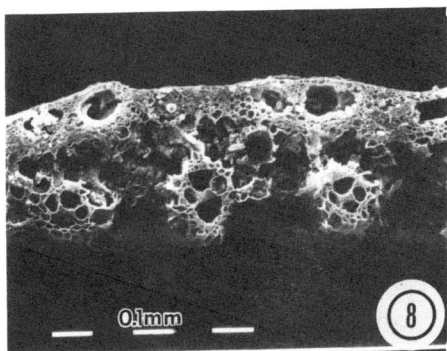

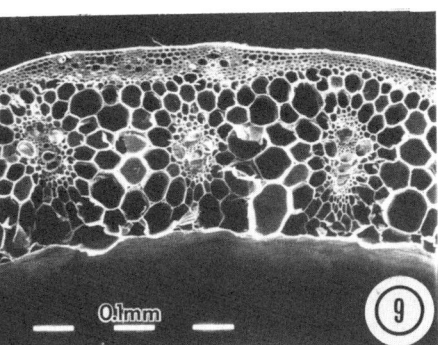
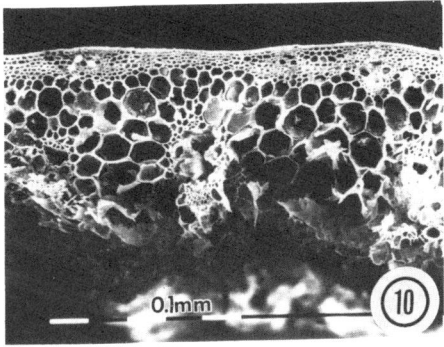

Plate 7. Transverse section of wheat stem at earing showing the epidermis, the sclerenchyma, and the vascular bundles in the parenchyma.   **Plate 8.** Wheat stem at earing after 24 h incubation in the rumen showing the extensive degradation of parenchyma and phloem tissues.   **Plate 9.** Transverse section of wheat steam at flowering showing the thickness of the sclerenchyma cell walls.   **Plate 10.** Wheat stem at flowering incubated 72 h in the rumen. A part of the parenchyma is undegradable (compare with Plate 8).

The study of tissue breakdown in teguments of soya bean and rape seed has been studied by SEM[2]. These two teguments are composed of several layers of cells. Soya bean tegument walls have almost no lignin whereas those of the external layer in rape seed are thick and highly lignified. After rumen digestion the soya bean teguments are almost entirely broken down (Plate 4) whereas the external layer of the rape seed remains intact (Plates 5,6).

Microscopic observation also makes it possible to follow changes in digestibility of different organs according to plant age. For example, the digestibility of stems is reduced more rapidly than that of leaves. This reduction results from changes in parenchyma walls, not lignified and very digestible in young plants but in which lignin is progressively layed down so that they become very indigestible in old plants. For wheat stems harvested every two weeks from earing to maturity, using LM we have shown the presence of lignin in walls by acid phloroglucinol stain and of phenolic compounds, precursors of lignin, by U.V. fluorescence. After incubation in the rumen of a fistulated cow, SEM showed that the walls of different stem tissues, with the exception of xylem, were highly digestible at earing stage (Plates 7,8)[16]. At flowering (Plate 9), cell walls were thicker and the sclerenchyma remained intact after 72 h incubation (Plate 10). Both lignified and the non-lignified parenchyma showed fluorescence in the U.V., indicating the presence of phenolic compounds in cell walls at all stages of maturity. However, as the plant ages, the cell walls of parenchyma become progressively lignified from the exterior of the stem towards the centre, demonstrated by acid phloroglucinol test and simultaneously become undegradable in the rumen. At grain maturity only the parenchyma fringing the medullary space is degraded and the fall in digestibility is associated with an increase in syringyl lignin which develops in sclerenchyma cell walls[17].

The effect of lignin in cell walls was first shown by Barton & Akin[18] with SEM, after delignification with potassium permanganate. Sclerenchyma walls were easier to treat than the bundle sheaths and xylem was not rendered digestible by this treatment. The results obtained show that cell wall polysaccharides differ considerably in their rate and extent of digestion when lignin no longer acts as a barrier.

Alkali treatment of straw is known to increase digestibility. This increase appears to be due to breakdown of linkages between cell wall carbohydrate and lignin. SEM observations of *Cynodon dactylon*[19] showed that cell walls were deformed after 10% potassium hydroxide treatment and that tissues were fragmented to separate cells. TEM studies confirmed this observation and showed, in addition, that bacteria become attached in greater number to the walls of treated samples, including the xylem[20]. The tissues are broken down slowly by adhering bacteria in the case of untreated samples, whereas they were rapidly degraded in treated samples. In addition, alkali treatment separated the xylem into individual cells and facilitated the degradation of the middle lamella of sclerenchyma by bacteria. In our observations of wheat straw treated with anhydrous ammonia (Grenet & Barry, unpublished), at least 72 h incubation in the rumen as necessary for the disorganisation of sclerenchya cells to occur and the beginning of their breakdown to become evident.

When tropical plants are subjected to drought, their cell walls and lignin contents increase whereas their crude protein and 48 h *in vitro* digestibility are reduced[21]. SEM observation of such plants, when compared with the same species grown under irrigation shows that their histological structure is identical

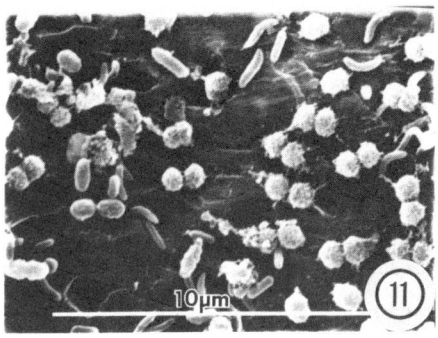

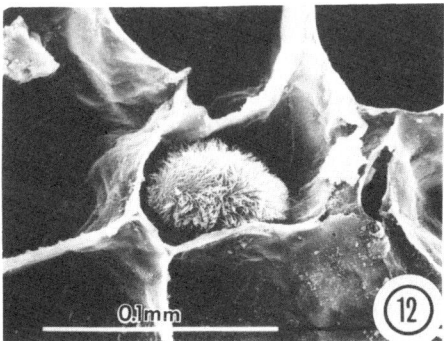
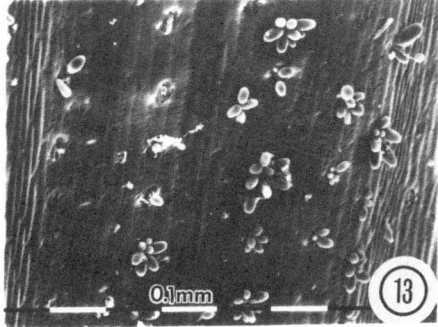

Plate 11. Two forms of rumen bacteria are shown adhering to wheat straw parenchyma cell walls (8 h incubation in the rumen). **Plate 12.** A protozoan is shown inside a parenchyma cell (sugar beet incubated 2 h in the rumen). **Plate 13.** Maize leaf incubated for 24 h in the rumen showing the colonisation of stomata by anaerobic fungi.

but that the number of cells with lignified walls is increased by drought stress, with the result that there is a reduced breakdown of the parenchyma bundle sheath in such plants.

Tropical species contain starch grains in the bundle sheath cells of leaves which may be observed by TEM. Wilson & Hattersley[22] observed this for *Panicum* and C_4 plants. With TEM, it has been demonstrated that these starch grains can be hydrolysed in the rumen only after cell wall breakdown. These walls contain a non-degradable suberin lamella which can only be broken down mechanically[23].

RELATIONS BETWEEN PLANT CELL WALLS AND RUMEN MICROORGANISMS

Bacteria

Observations of plant fragments incubated *in vitro* with rumen juice, with pure cultures of microorganisms or after passage in the rumen, show the presence of adherent bacteria. Beet pulp placed in the rumen in nylon bags for 8 h becomes covered with a whitish film caused by the presence of numerous bacteria attached to cell walls[24]. SEM permits identification of two forms: cocci and rods, similar to those observed on wheat straw incubated in the rumen for the same time (Plate 11). A filamentous microorganism was observed by SEM and TEM by Akin[25] fixed to plant tissues which it was able to degrade. Similarly, a facultative anaerobic bacterium was isolated in the rumen and its role in *in vivo* degradation of plant cell walls studied by SEM[26]. *In vitro* studies by Akin & Rigsby[(27)] using SEM, provided observation on the disappearance of plant tissues due to action of a single bacterial species. Fixation of *Ruminococcus flavefaciens in vitro*[28] on damaged sections of plant tissues was observed by SEM, together with the presence of chains of cocci along epidermal walls, at the cuticular edge. These microorganisms gave rise to cavities in plant tissues. *Bacteroides succinogenes* also attached to sectioned edges of cell walls of all tissues except xylem but was found in smaller quantities on the intact surfaces of plant tissues[29]. In the SEM study made by Fay *et al.*[30] of the bacterial colonisation of forage legumes, sainfoin, which does not generate bloat was colonized more slowly than lucerne, which may induce bloat.

Although certain bacteria attach only to certain forage types, there is no relation known between forage digestibility and the type of bacteria[31]. In general, bacteria become most closely attached to cell walls which are slow or difficult to degrade and are less closely associated with the easily degradable cell walls, such as the primary walls of mesophyll, epidermis or phloem.

When examining the digestion of legume leaves by TEM, Cheng *et al.*[32] observed adhesion of bacteria to cell walls, their proliferation around stomata, followed by their development in the intercellular spaces in the form of microcolonies enclosed in exopolysaccharides, and the formation of intracellular colonies.

Direct observation of cell wall breakdown by cellulolytic bacteria shows that they adhere to the surface of cell walls through their glycocalyx. If conditions are favourable, the bacteria multiply and form microcolonies on the cell wall surface[20]. Cell walls with different chemical compositions are colonized by different types of cellulolytic bacteria[29]. Legume cell walls rich in pectin are colonised by *Lachnospira*[33] and by cocci resembling *Ruminococcus*[34], whilst the primary walls of Gramineae are mainly colonised by *Ruminococcus flavefaciens*[35] and straw cell walls by *Bacteroides succinogenes*[36]. The latter adhere very closely to the cellulosic substrate by means of their glycocalyx and they continue to digest the substrate by formation of vesicles presumed to contain polysaccharidase enzymes, which function even after autolysis of the cell[37].

Protozoa

SEM observation shows the presence of considerable numbers of protozoa on the plant substrates. Entodiniomorph protozoa are found in large numbers on beet pulp placed in the rumen for 4 h in nylon bags. Holotrichs protozoa are found within beet parenchyma cells that have spent 2 h in the rumen (Plate 12). The cell contents, rich in sugars are utilized by these protozoa. *Epidinium* is found also in the mesophyll of maize leaves placed in the rumen for 4 h. These microorganisms can breakdown large quantities of mesophyll, fragments of parenchyma bundle sheaths and epidermis of temperate Gramineae leaves[38] in addition to fragments of lucerne stems[39,40]. Observation with the TEM shows this breakdown appears to result from the action of both extracellular enzymes and also the ingestion of fragments of cells and cell walls[41,42].

Fungi

SEM observations of the breakdown of plant cell walls have demonstrated the presence of many sporangia of anaerobic rumen fungi on certain plant substrates placed in the rumen in nylon bags. They are more numerous on stems than on leaves, except when the leaves are not very digestible (straw or *Agave sisalona*). The spores fix on lucerne stems, within grass stems and around stomata (Plate 13). It is particularly interesting to note that they are found fixed on the sclerenchyma of grass leaves[43,44] and, in general, on tissues with thick or lignified walls[45]. In contrast other plant substrates such as beet pulp and rape seed tegument are only rarely colonized. Shortly (15 min) after the introduction into the rumen of nylon bags containing plant fragments, fixation of fungal spores can be observed[46]. They germinate to give rise to a mycelium which can be observed 3 h after spore fixation. The rhizoids of mycelium penetrate plant tissue to a depth which may reach 460 µm. Using SEM, Ho *et al.*,[47] observed the presence of fungal rhizoids with fingerlike appressoria which enabled them to penetrate cell walls. SEM observations have also shown that the animals' diet has an effect on the development of rumen fungi[48]. Fungi are particularly abundant with diets rich in forage but are fewer in number in animals fed starch or soluble carbohydrate rich diets. The reduction in the number of spores in the rumen juice is accompanied by a less abundant colonisation of plant substrates and reduced development of sporangia[48].

The role of anaerobic fungi in rumen degradation is only now beginning to be understood[49]. Microscopic observations have shown that part of lignified tissue is degraded *in vitro* by fungal cultures[43] in particular sclerenchyma. SEM observations have been made recently at Theix[50] on the breakdown of plant tissues in the rumen of gnotoxenic lambs with fungi of the genus *Neocallimastix* as the only cellulolytic microorganism (Fonty, unpublished). Fragments of

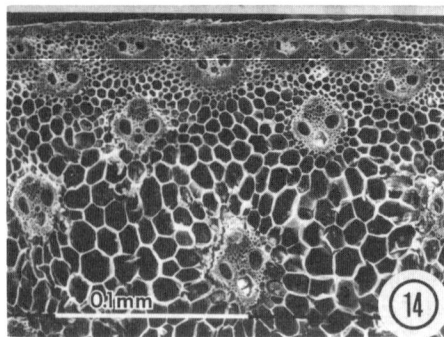 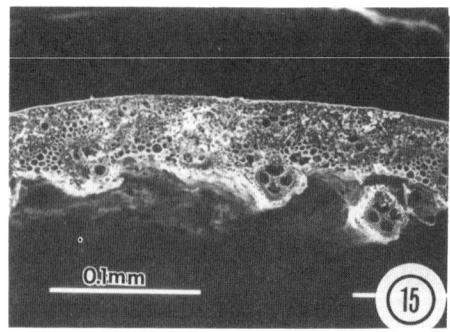

Plate 14. Maize stem incubated 8 h in the rumen of gnotoxenic lambs containing *Neocallimastix* sp. **Plate 15**. Maize stem incubated 48 h in the rumen of gnotoxenic lambs containing *Neocallimastix* sp. showing the loss of parenchyma and the phloem tissue.

maize stem, for instance, have been placed, in nylon bags, in the rumen for 8, 24, 48 and 72 h. If the results obtained are compared with those obtained after equivalent stay of the same material in conventional rumen, it is found that tissue breakdown is slower, but after 48 h the results are similar. The lignified tissues are highly colonized in the rumen of the gnotoxenic lambs but it is the cellulosic tissues (parenchyma, phloem) which are degraded (Plates 14, 15).

The ultrastructure of anaerobic rumen fungi have been studied by TEM. The zoospores of *Neocallimastix* are uninucleate and contain numerous inclusions[51]. Close to the flagellary pole are localized microbodies surrounded by a membrane, particles, small vesicles and microtubules which, in groups of about 14, diverge from the kinetosome of each flagellum. The microbodies are the base of enzymatic activities involving a hydrogenase and thus may be equivalent to hydrogenosomes[52]. There are no mitochondria.

CONCLUSION

Electron microscopy makes it possible to study the nature of tissues which make up lignocellulosic feeds and the ultrastructure of the cell walls in these tissues. This technique has permitted demonstration of the wide variability that exists in the capacity for degrading plant tissues and their selective breakdown in the rumen. It has provided an understanding of the mechanisms which cause reduction in digestibility with plant age, how the presence of lignin or silica affect cell wall breakdown and the effects of chemical treatments on degradation. The relationship between plant cell walls and microorganisms is particularly well demonstrated with these techniques, especially the invasion of tissues by microbes and their selective attachment to cell walls.

Statistical analysis of the results obtained by electron microscopy remains difficult at present since numerical values are generally difficult to obtain. The areas observed must be selected with care to provide representative observations and different samples must be studied to provide a representation of the whole plant.

The future development of microscopic studies will be based on inmunocytochemical techniques. These techniques already permit the localisation in cell walls of xylans by an enzymatic method with gold labelling[53,54]. Unfortunately, it is not yet possible to locate all of the constituents of the cell wall, nor to demonstrate the presence of polysaccharide-lignin linkages which are the touchstone of plant cell wall breakdown. However, inmunocytochemical methods, using monoclonal antibodies, should make it possible to determine with precision the internal localisation of bacterial secretion of enzymes such as cellulases (endoglucanase, cellobiase). Research on this topic is in progress in the laboratory of Dr. Forsberg, Canada (Cheng, personal communication). The information obtained should provide a better understanding of the mode of activity of microorganisms. These techniques should also make it possible to demonstrate the enzymatic secretion of plant during development of their cell walls.

Microscopic observations can be complementary to studies characterising plant structure by ^{13}C-NMR spectroscopy or HPLC. The two methods make it possible to determine the structure of lignin-carbohydrate complexes and their effects on digestion. Further, solid-state NMR spectroscopy is a new technique which should provide additional information about cell wall structure. Finally, IR and NIR spectroscopy are also modern, efficient techniques which can be used to predict the quality of forages.

Microscopy is a highly specialized technique which gives detailed information on the mechanisms of microbial degradation. It is an important addition to studies using other physcal or chemical methods on the breakdown of plant cell walls in the understanding of the functioning of the rumen and from this, improvement in ruminant nutrition.

REFERENCES

1.   Grenet, E. and Barry, P. (1988). Dégradation microbienne dans le rumen du maïs "brown midrib" observée au microscope électronique á balayage. *Reproduction Nutrition Développement* **28**: 125-126.

2.   Grenet, E. and Barry, P. (1987). Etude microscopique de la digestion des parois végétales des téguments de soja et de colza dans le rumen. *Reproduction Nutrition Développement*, **27**: 246-248.

3.   Akin, D.E. and Burdick, D. (1973). Microanatomical differences of warm-season grasses revealed by light and electron microscopy. *Agronomy Journal* **65**: 533.

4. Harbers, L.H.; Brazle, F.K.; Raiten, D.J. and Owensby, C.E. (1980). Microbial degradation of smooth brome and tall fescue observed by scanning electron microscopy. *Journal of Animal Science* **51**: 439-446.

5. Harbers, L.H.; Raiten, D.J. and Paulsen, G.M. (1981). The role of plant epidermal silica as a structural inhibitor of rumen microbial digestion in steers. *Nutrition Reports International.* **24**: 1057-1066.

6. Shiro Saka, Thomas, J.R. and Gratzl, J.S. (1984). Lignin distribution by energy disposal. In *Dietary fibres, Chemistry and Nutrition.* Inglett, G.E. and Falkehag, S.I. (eds). pp. 15-29. Academic Press: New York.

7. Czaninski, Y. (1979). Cytochimie ultrastructurale des parois du xyléme secondaire. *Biology of the Cell* **35**: 97-102.

8. Thiery, J.P. (1967). Mise évidence des polysaccharides sur coupes fines en microscopie électronique. *Journal of Microscopy* **6**: 987-1018.

9. Luft, J.H. (1971). Ruthenium red and violet. I - Chemistry, purification, methods of use for electron microscopy and mechanism of action. *Anatomical Record* **171**: 347-368.

10. Albersheim, P. (1965). A cytoplasmic component stained by hydroxylamine and iron. *Protoplasma.* **60**: 131-135.

11. Coppick, S. and Fowler, W.F. (Jr.) (1939). The location of potential reducing substances in woody tissues. *Paper Trade Journal* 81-86.

12. Engels, F.M. and Brice, R.E. (1985). A barrier covering lignified cell walls of barley straw that restricts access by rumen microorganisms. *Current Microbiology.* **12**: 217-224.

13. Akin, D.E. and Burdick D. (1975). Percentage of tissue types in tropical and temperate grass leaf blades and degrdation of tissues by rumen microorganisms. *Crop Science* **15**: 661.

14. Akin, D.E. and Burdick, D. (1981). Relationships of different histochemical types of lignified cell walls to forage digestibility. *Crop Science* **21**: 577-581.

15. Akin, D.E. and Hogan, J.P. (1983). Sulphur fertilization and rumen microbial degradation of cell walls in *Digitaria pentzii* (Stent). *Crop. Science* **23**: 851-858.

16. Barry, P. and Grenet, E. (1988). Dégradation microbienne dans le rumen de la tige de blé á différents stades de développement, observée au microscope électronique á balayage. *Reproduction Nutrition Dévelopement* (in press).

17. Akin, D.E.; Robinson, E.L.; Barton II, F.E. and Himmelsbach, D.S. (1977). Changes with maturity in anatomy, histochemistry, chemistry and tissue digestibility of bermudagrass plant parts. *Journal of Agricultural and Food Chemistry* **25**: 179.

18. Barton, II, F.E. and Akin, D.E. (1977). Digestibility of delignified forage cell walls. *Journal of Agricultural and Food Chemistry* **25:** 1299.

19. Spencer, R.R. and Akin, D.E. (1980). Rumen microbial degradation of potassium hydroxide-treated coastal bermudagrass leaf blades examined by electron microscopy. *Journal of Animal Science* **51:** 1189-1196.

20. Cheng, K.J.; Stewart, C.S.; Dinsdale, D. and Costerton, J.W. (1984). Electron microscopy of bacteria involved in the digestion of plant cell walls. *Animal Feed and Science Technology* **10:** 93-120.

21. Akin, D.E.; Barton, II, F.E. and Coleman, S.W. (1983). Structural factors affecting leaf degradation of old world bluestem and weeping love grass. *Journal of Animal Science* **56:** 1434-1446.

22. Wilson, J.R. and Hattersley, P.W. (1983). *In vitro* digestion of bundle sheath cells in rumen fluid and its relation to the suberized lamella and C_4 photosynthetic type in *Panicum* species. *Grass Forage Science* **38:** 219-223.

23. Hastert, A.A.; Owensby, C.E. and Harbers, L.H. (1983). Rumen microbial degradation of indiangrass and big bluestem leaf blades. *Journal of Animal Science* **57:** 1626-1636.

24. Grenet, E. and Barry, P. (1986). Rumen microbial degradation of soyabean hulls, rape seed hulls and dehydrated beet pulp examined by scanning electron microscopy. Comparative aspects of physiology of digestion in ruminants. A satellite meeting of the XXX International Congress of the International Union of Physiological Sciences. Cornell University, July 21-29. pp. 16-17.

25. Akin, D.E. (1976). Ultrastructure of rumen bacterial attachment to forage cell walls. *Applied and Environmental Microbiology* **31:** 562.

26. Akin, D.E. (1980). Attack on lignified grass cell walls by a facultative anaerobic bacterium. *Applied and Environmental Microbiology* **40:** 809-820.

27. Akin, D.E. and Rigsby, L.L. (1985). Degradation of bermuda and orchard grass by species of ruminal bacteria. *Applied and Environmental Microbiology* **50:** 825-830.

28. Latham, M.J.; Brooker, B.E.; Pettipher, G.L. and Harris, P.J. (1978a). *Ruminococcus flavefaciens* cell coat and adhesion to cotton cellulose and cell walls in leaves of perennial ryegrass (*Lolium perenne*). *Applied and Environmental Microbiology* **35:** 156-165.

29. Latham, M.J.; Brooker, B.E.; Pettipher, G.L. and Harris, P.J. (1978b). Adhesion of *Bacteroides succinogenes* in pure culture and in the presence of *Ruminococcus flavefaciens* to cell walls in leaves of perennial ryegrass (*Lolium perenne*). *Applied and Environmental Microbiology* **35:** 1166-1173.

30. Fay, J.P.; Cheng, K.J. and Hanna, M.R. (1981). A scanning electron microscopy study of the invasion of leaflets of a bloat-safe and a bloat-causing legume by rumen microorganisms. *Canadian Journal of Microbiology* **27:** 390-399.

31. Akin, D.E. and Barton, II, F.E. (1983). Rumen microbial attachment and degradation of plant cell walls. *Federation Proceedings* **42:** 114-121.

32. Cheng, K.J.; Fay, J.P.; Howarth, R.E. and Costerton, J.W. (1980). Sequence of events in the digestion of fresh legume leaves by rumen bacteria. *Applied and Environmental Microbiology* **41:** 298-305.

33. Cheng, K.J.; Dinsdale, D. and Stewart, C.S. (1979). The maceration of clover and grass leaves by *Lachnospira multiparus*. *Applied and Environmental Microbiology* **38:** 723-729.

34. Cheng, K.J.; Brown, R.G. and Costerton, J.W. (1977). Characterization of a cytoplasmic reserve glucan from *Ruminocuccus albus*. *Applied and Environmental Microbiology* **33:** 718-724.

35. Dinsdale, D.; Morris, E.J. and Bacon, J.S.D. (1978). Electron microscopy of the microbial populations present and their modes of attachment on various cellulosic substrates undergoing digestion in the sheep rumen. *Applied and Environmental Microbiology* **36:** 160-168.

36. Stewart, C.S.; Dinsdale, D.; Cheng, K.J. and Paniagua, C. (1979). The digestion of straw in the rumen. In *Straw decay and its effect on disposal and utilization*. E. Grossbard (ed.) p. 123-130 Wiley Interscience.

37. Chesson, A.; Stewart, C.S., Dalgarno, K. and King, T.P. (1986). Degradation of isolated grass mesophyll, epidermis and fibre cell walls in the rumen and by cellulolytic rumen bacteria in axenic culture. *Journal of Applied Bacteriology* **60:** 327-336.

38. Amos, H.E. and Akin, D.E. (1978). Rumen protozoal degradation of structurally intact forage tissues. *Applied and Environmental Microbiology* **36:** 513-522.

39. Senaud, J.; Bohatier, J. and Grain, J. (1986). Comparaison du comportement alimentaire de 2 Protozoaires Ciliés du rumen, *Epidinium caudatum, Polyplastron multivesiculatum*: mécanismes d'ingestion et de digestion. *Reproduction Nutrition Développement* **26:** 287-289.

40. Bauchop, T. (1979). The rumen ciliate *Epidinium* in primary degradation of plant tissues. *Applied and Environmental Microbiology* **37:** 1217-1223.

41. Akin, D.E. and Amos, H.E. (1979). Mode of attack on orchard grass leaf blades by rumen protozoa. *Applied and Environmental Microbiology* **37:** 332-338.

42. Grain, J. and Semaud, J. (1984). New data on the degradation of fresh lucern fragments by the rumen ciliate *Epidinum ecaudatum*: attachment, ingestion and digestion. *Canadian Journal of Animal Science* **64**: 26.

43. Akin, D.E.; Gordon, G.L.R. and Hogan, J.P. (1983). Rumen bacterial and fungal degradation of *Digitaria pentzii* grown with or without sulfur. *Applied and Environmental Microbiology* **46**: 738-748.

44. Akin, D.E. (1987). Association of rumen fungi with various forage grasses. *Animal Feed Science and Technology* **16**: 273-285.

45. Grenet, E. and Barry, P. (1987). Colonization of thick walled tissues by anaerobic rumen fungi. *Animal Feed Science and Technology* **19**: 25-31.

46. Bauchop. T. (1984). Rumen anaerobic fungi and the utilisation of fibrous feeds. In *Biotechnology and Recombinant DNA Technology in the Animal Production Industries,* Reviews in Rural Science 6. p. 118-123 (Leng, R.A., Barker, J.S.F., Adam, D.B. and Hutchinson, K.L. Eds). *Symp. Univ. New Engl.* Armidale, Australia.

47. Ho, Y.W.; Abdullah, N. and Jalaludin, S. (1988). Penetrating structures of anaerobic rumen fungi in cattle and swamp buffalo. *Journal of General Microbiology* **134**: 117-181.

48. Grenet, E.; Breton, A.; Barry, P. and Fonty, G. (1989). Rumen anaerobic fungi and plant substrate colonization as affected by diet composition. *Animal Feed Science and Technology* (in press).

49. Fonty, G.; Grenet, E.; Fevre, M.; Breton, A. and Gouet Ph. (1989). Biologie et fonctions des champignons anaérobies du rumen. *Reproduction Nutrition Développement* (in press).

50 Grenet, E.; Fonty, G. and Barry, P. (1988). Study of the degradation of maize and lucerne stems in the rumen of gnotobiotic lambs harbouring only fungi as cellulolytic microorganisms. Proceedings of the International Seminar on "The role of protozoa and fungi in ruminant digestion". Armidale, Australia (in press).

51. Munn, E.A.; Orpin, C.G. and Hall, F.J. (1981). Ultrastructural studies of the free zoospore of the rumen phycomycete *Neocallimastix frontalis*. *Journal of General Microbiology* **125**: 311-323.

52. Yarlett, N.; Orpin, C.G.; Munn, E.A.; Yarlett, N.C. and Greenwood, C.A. (1986). Hydrogenosomes in the rumen fungus *Neocallimastix partriciarum*. *Biochemical Journal* **236**: 729-739.

53. Vian, B.; Brillouet, J.M. and Satiat-Jeunemaitre, B. (1983). Ultrastructural visualization of xylans in cell walls of hardwood by means of xylanase-gold complex. *Biology of the Cell* **49**: 179-182.

54. Ruel, K. and Joseleau, J.P. (1984). Use of enzyme-gold complexes for the ultrastructural localization of hemicelluloses in the plant cell wall. *Histochemistry* **81**: 573-580.

SOME PROPERTIES OF CELL WALL LAYERS DETERMINING RUMINANT DIGESTION

F.M. ENGELS

Department of Plant Cytology and Morphology, Agricultural University Wageningen, Arboretumlaan 4, 6703 BD Wageningen, The Netherlands

SUMMARY

The digestion of cell walls of different grasses was studied using light- and electron microscopy. A new technique was developed which enables sections of plant tissues containing exactly the same cell walls to be used in different experiments and treatments. Using this method it was observed that cell wall digestion was limited by a specific cell wall region, common to all grasses investigated, which probably represented the middle lamella-primary cell wall. Strongly phloroglucinol staining (lignified) secondary cell wall layers were, however, digested. Delignification with potassium permanganate was not successful for the middle lamella-primary wall. It is suggested that low amounts of permanganate-resistant lignin deposited in the middle lamella-primary wall region during plant growth acts to restrict the rumen digestion of plant tissues.

INTRODUCTION

During the last ten years light and electron microscopic techniques have provided data which allows a better understanding of the mechanism underlying the decreasing digestibility of plant cell walls observed when plants grow older. Studies of cell wall residues recovered after rumen digestion by microorganisms have suggested the presence of a number of factors which function as barriers to the progress of microbial digestion. These include the epidermal cuticle,[10] the tertiary wall[3], a warty layer[6], the lignified primary cell wall,[12] suberised lamella[14] and a lignin enriched layer which forms at cell wall surfaces during degradation.[5] In each case it is believed that the non-carbohydrate moiety of the cell wall (e.g. cutin, suberin, lignin) is the major factor inhibiting digestion, because of its chemical linkage to the cell-wall carbohydrates.

CHARACTERIZATION OF CELL WALL STRUCTURE

Electron micrographs made from sections of digested barley straw provide evidence of a cell wall residue which derives from two originally adjacent cell walls. Examination of sections from undigested controls strongly suggest that this cell wall residue has its origin in the middle lamella region of the primary wall (mlpw, Engels & Brice, unpublished results). Examples of such a cell

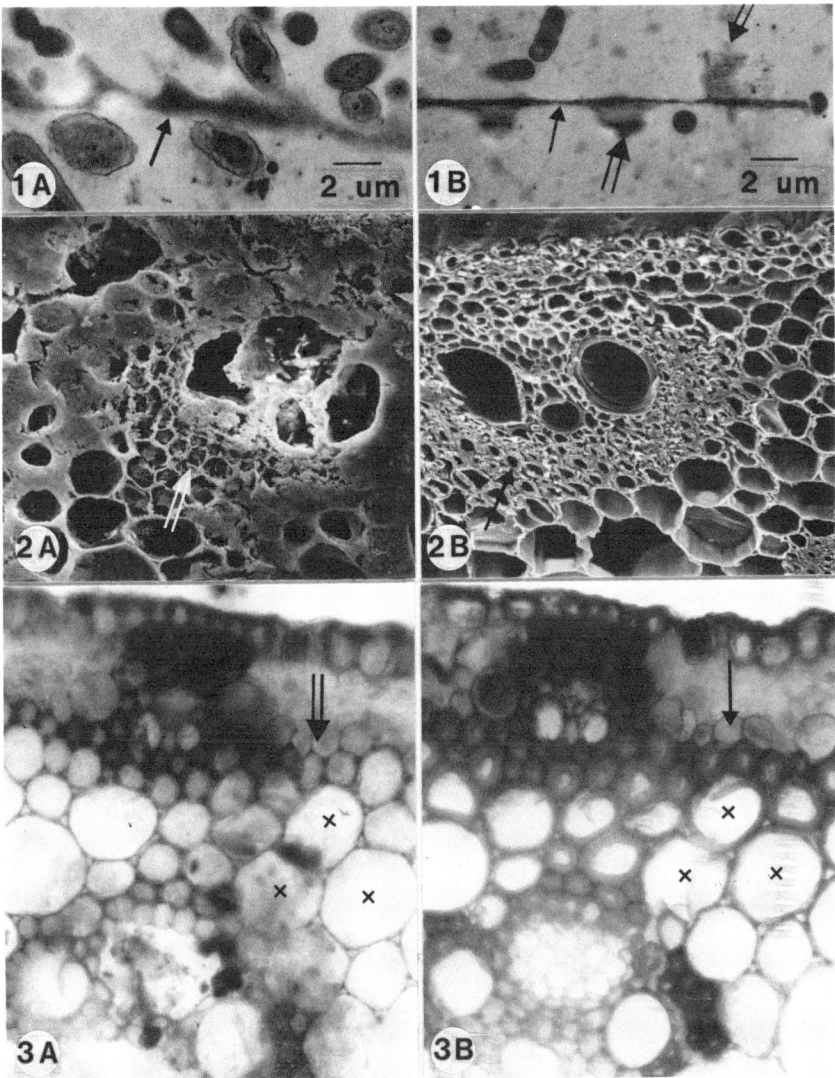

Plate 1. TEM of digested barley parenchyma cell walls. (A) Dark stained cell wall residue (arrow) with some adherent bacteria. (B) Blocks of non-digested cell wall which indicate the original cell wall thickness (wide arrow). **Plate 2.** SEM of maize stem sections. (A) A thin cell wall residue originating from sclerenchyma is found after digestion (arrow). (B) Undigested sclerenchyma showing the thick cell walls found surrounding the vessel (arrow). **Plate 3.** Light micrograph of *Panicum* sp. before and after digestion. The thick walled sclerenchyma tissue (arrow, B) is reduced to very thin cell wall residue during digestion (wide arrow, A). In subsequent 100 μm sections the same cells can be recognised because of the large size of some individual cells (x).

wall residue (from parenchyma) are shown in Plate 1. The less intensively stained cell walls are broken down leaving behind a dark stained, nearly complete, residue probably representing the mlpw.

In a study of maize (var. LG11) it was found that large parts of the sclerenchyma cell walls could be digested (Cone & Engels, in preparation). However, the tissue did not loose its anatomical and morphological organisation during microbial degradation and, again, the mlpw's appeared to form the bulk of the cell wall residue. This is exemplified in Plate 2 in which digested (Plate 2A) and non-digested (Plate B) stem sections are compared. In the tropical grass *Panicum maximum* the same type of digestion of cell walls was observed (Engels & Wilson, unpublished results) as shown in Plate 3. It is evident from Plate 3A that sclerenchyma cell walls are considerably reduced in thickness during incubation in rumen fluid.

Maize stems (var. Brutus) from plants grown under field conditions were collected at the time of pollen shedding. The stem internode bearing the kernel (5th or 6th internode from top) was selected for study. Thin 100 μm sections cut at different levels were stained with 1% phloroglucinol in 20% HCl (AP) for 12 minutes. Figure 1 summarises the results obtained from observations made over the entire internode.

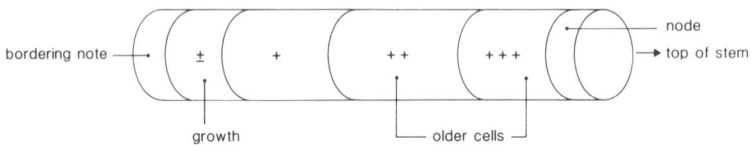

Figure 1. A diagrammatic representation of a maize stem internode, showing the intensity of staining of sections with AP (see text for details). ±, weakly staining; +++, intense staining reaction.

Internode growth occurs at the base of the internode close to the lower bordering node. Cells in this region were weakly stained while it was observed that the cell walls from the more mature part of the internode reacted more intensively with AP (Figure 1).

To enable the study of walls from the same cell under various conditions a mirror sectioning technique (ms) of the stem was developed. This is shown schematically in Figure 2.

Section A is turned 180° before it is fixed onto a double-sided adhesive tape and is thus the mirror image of section B. Two different treatments can be applied to exactly the same cell walls and can be studied in details when the colour slide positives are compared carefully one above the other (e.g. section

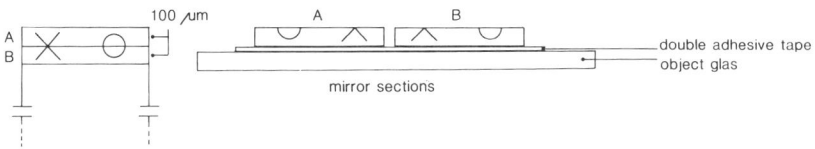

Figure 2. Method of preparing mirror sections of plant (maize) stems

A non-treated, and section B digested and AP stained). This technique has proved of considerable value in determining the relative degradability of cell wall parts in relation to their staining reaction.

Plate 4 (a ms set) shows a sclerenchyma bundle from maize before (Plate 4A) and after digestion in rumen fluid (Plate 4B). After digestion it is obvious that the thickness of the cell wall has been reduced considerably. The thickness of two adjacent cell walls was 2.7 µm, which was reduced to 0.36 µm after digestion. Staining of the section shown in Plate 4A gave a very intensive red AP reaction throughout the tissue. Strikingly the residual cell walls (Plate 4B) were found to have approximately the same thickness as found previously for isolated primary cell walls from leaf mesophyll cells.[9] This is a further indication that only the secondary wall was degraded, leaving the modified primary wall largely intact. Underneath the epidermis in maize stems, the vessel bundles are surrounded by a thick sclerenchyma sheet (Plate 5A) from which all the secondary cell wall can be digested (Plate 5B). Remarkably, more residual cell wall material was found at the cell corners. In the same ms set (Plate 5) the highly positive AP reaction of cell wall material of undigested (Plate 5C) and digested (Plate 5D) tissues is shown. These results demonstrate that the AP staining can only be used to detect phenolic compounds rather than to predict digestibility. This is in contrast to the general conception that tissues which stain for lignin with AP are resistant to ruminal digestion. Very recently Akin (this meeting) and Garcia & Latg'e[17] have identified a range of simple and complex phenolic compounds that give a positive colour reaction with AP.

DELIGNIFICATION BY KMnO₄ AND AP STAINING

It is well known that $KMnO_4$ oxidation can be used to delignify plant material. Maize sections were treated with saturated $KMnO_4$ for at least 18 hrs and subsequently with oxalic acid dihydrate until they become colourless (modified after Goering & van Soest[1]). The sections were subsequently stained with AP. This method is simple but the sections must be handled very carefully because they are very soft and easy to disturb. This treatment results in a colourless secondary cell wall and a soft yellow-orange stained mlpw (arrows, Plate 6). The colour positive material is identical with the cell wall residue found after digestion (Plate 7). From these observations we conclude that (i)

84

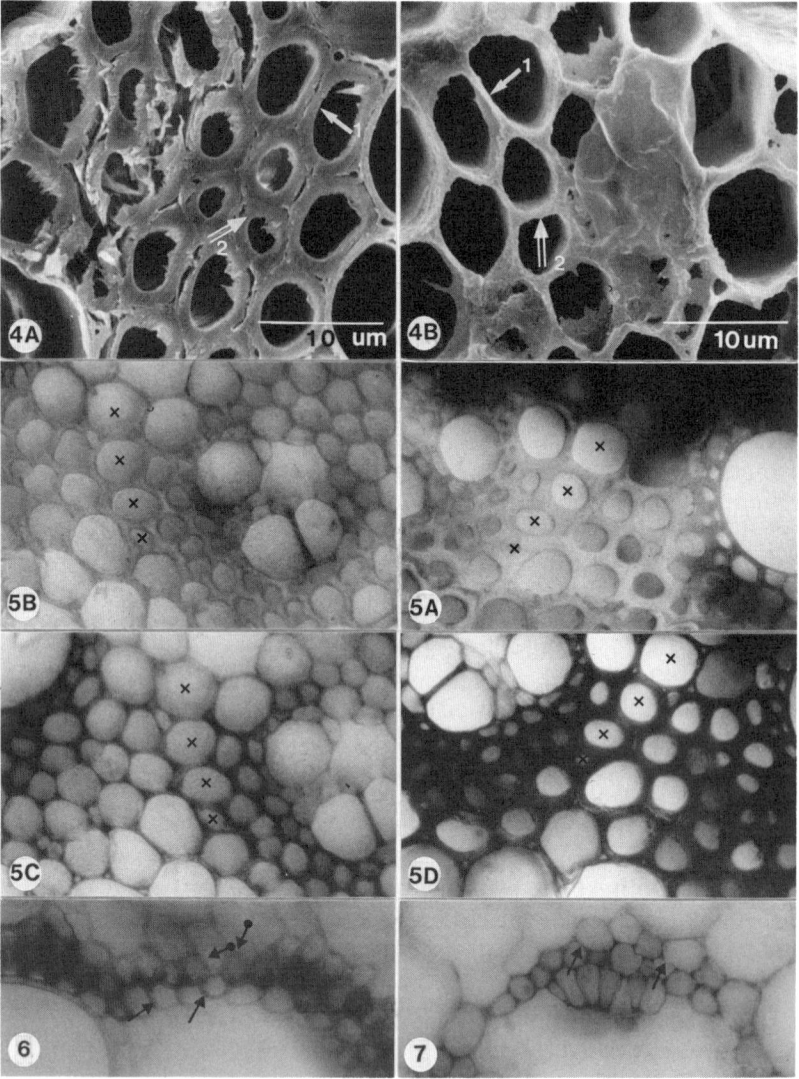

Plate 4. Mirror set of sections of maize sclerenchyma. The very thick cell walls seen in (A) are reduced considerably during digestion (B). The arrows point to exactly the same cell walls. **Plate 5.** Thick wall sclerenchyma tissue of maize stem. Each cross (x) marks exactly the same cell and cell wall obtained by the mirror sectioning technique. (A) Nondigested, unstained. (B) Digested, unstained. (C) as (A) but stained with AP. (D) as (B) but stained with AP. It is evident that only a small part of the AP-postive cell wall is highly resistant to digestion by rumen microorganisms. **Plate 6.** KMnO₄ delignification and AP staining of maize cells results in a colourless secondary cell wall (dotted arrow) while the thin mlpw's (arrow) stained positive with AP. **Plate 7.** Digested section of maize stem stained with AP. During digestion the secondary cell wall is removed and a very thin mlpw is left (arrow) which stains intensively with AP.

incrustation of $KMnO_4$ resistant material occurs in the mlpw surrounding the living cell, (ii) the presence of this material in the mlpw turns it into an indigestible layer, (iii) from the two adjacent cell walls separated by a mlpw, only the secondary wall in a mechanically opened cell lumen can be digested. However it is important to stress that experiments described here were carried out with relatively young material. Other experiments made with older material (roughage maize) gave similar results but some thickening of the $KMnO_4$ resistant layer was evident.

Akin *et al.*[2] found AP reacting cell walls throughout Bermuda grass after prolonged $KMnO_4$ oxidation unlike the results obtained with maize stems. It is possible that incrustation with $KMnO_4$-resistant material, as found in maize, could occur much earlier and to a greater extent throughout the entire cell wall in Bermuda grass, this making the wall completely indigestible.

ADDITIONAL OBSERVATIONS AND CONCLUSION

Parenchymatous tissue from maize stem is fully digested *in vitro* provided young tissue is used. After a certain age during growth of the plant degradation of parenchyma is much reduced as was found for sclerenchyma cell walls. We found large amounts of parenchyma with only very superficial cavities on the cell wall produced by bacterial activity. Although the cell walls of two adjacent cells are very thin (± 0.3 μm) it was observed that this tissue remained unaltered even after 140 hr of *in vitro* digestion. In a 1 cm stem section only 2-3 cell layers located at the cut end were attacked by microorganisms, representing only about 100 μm of the entire 1 cm length. In the author's opinion continuous mechanical disruption is required to allow cell wall breakdown to progress because the mlpw will not be digested if $KMnO_4$-resistant material is present. The role of fungi in fractionation of larger particles to smaller ones seems rather doubtful because we have observed morphologically perfectly intact tissues covered with an abundance of fungal sporangia. Comparable results have been reported elsewhere.[4,11,15]

Our findings with maize leads to the conclusion that (i) there is an accumulation of phenolic compounds in the secondary wall which react positively with AP, but which does not impede cell wall digestibility, (ii) there is an incrustation of $KMnO_4$-resistant material which reacts positively with AP and which forms an effective barrier to digestion even found in low amounts in the mlpw, (iii) if this material is also impregnated in the secondary wall, this part of the wall will also become indigestible, (iv) in older plants more mlpw's react positively with AP after $KMnO_4$ treatment, an observation that can be correlated with a decreasing availability of secondary cell walls for digestion. Therefore one has to be very careful in directly relating amounts of carbohydrates and lignin present with the degree of digestibility.

Further research is aimed at the isolation of mlpw's without any secondary cell wall for chemical analysis. Parts of this paper will be published elsewhere in greater detail. This work was made possible in close cooperation with Drs. R.E. Brice (United Kingdom), J.W. Cone (Netherlands) and J.R. Wilson (Australia).

REFERENCES

1. Akin, D.E. and Burdick, K.D. (1981). Relationships of different histochemical types of lignified cell walls to forage digestibility. *Crop Science* **21:** 577-581.

2. Akin, D.E.; Willemse, M.T.M. and Barton, F.E. (1985). Histochemical reactions, autofluorescence, and rumen microbial degradation of tissues in untreated and delignified Bermuda grass stems. *Crop Science* **25:** 901-905

3. Bacon, J.S.D. (1979). Plant cell wall digestibility and chemical structure. *Report of the Rowett Institute* **35:** 99-108.

4. Calderon-Cortes, J.F. (1988). Influence of feed supplements on rumen microbial activity. Ph.D. Thesis. University of Queensland.

5. Chesson, A. (1988). Lignin polysaccharide complexes of the plant cell wall and their effect on microbial degradation in the rumen. *Animal Feed Science and Technology* **21:** 219-228.

6. Engels, F.M. and Brice, R.E. (1985). A barrier covering lignified cell walls of barley straw that restricts access by rumen microorganisms. *Current Microbiology* **12:** 217-224.

7. Garcia, S. and Latgé, J.P. (1987). A new colorimetric method for dosage of lignin. *Biotechnology Techniques* **1:** 63-68.

8. Goering, K.H. and van Soest, P.J. (1970). *Forage fibre analyses* USDA Agricultural Handbook 379. 1-20.

9. Gordon, A.H.; Lomax, J.A.; Dalgarno, K. and Chesson, A. (1985). Preparation and composition of mesophyll, epidermis and fibre cell walls from leaves of Perennial rye grass (*Lolium perenne*) and Italian rye grass (*Lolium multiflorum*). *Journal of the Science of Food and Agriculture* **36:** 509-519.

10. Latham, M.J.; Brooker, B.E.; Pettipher, G.L. and Harris, P.J. (1978). Adhesion of *Bacteroides succinogenes* in pure culture and in the presence of *Ruminococcus flavefaciens* to cell walls in leaves of Perennial rye grass (*Lolium perenne*). *Applied and Environmental Microbiology* **35:** 1166-1173.

11. McLeod, M.N. and Minson, D.J. (1988). Large particle breakdown by cattle eating ryegrass and alfalfa. *Journal of Applied Science* **66:** 992-999

12. Ride, J.P. (1983). Cell walls and other structural barriers in defence. In *Biochemical Plant Pathology*. J.A. Callow (ed.) John Wiley and Sons Ltd.

13. Schmidt, O. and Liese, W. (1982). Bacterial decomposition of woody cell walls. *International Journal of Wood Preservation* **2:** 13-19.

14. Wilson, J.R. and Hattersley, P.W. (1983). *In vitro* digestion of bundle sheath cells in rumen fluid and its relation to the suberized lamella and C4 photosynthetic type in *Panicum* species. *Grass and Forage Science* **38:** 219-223.

15. Wilson, J.R. and Engels, F.M. (1988). Do rumen fungi have a significant direct role in particle size reduction? Intern. Symp. The rule of protozoa and fungi in ruminant digestion. Armidale Australia (in press).

IDENTIFICATION AND IMPORTANCE OF POLYPHENOLIC COMPOUNDS IN CROP RESIDUES

I. MUELLER-HARVEY

AFRC Institute for Grassland and Animal Production, Animal and Grassland Research Station, Hurley, Maidenhead, SL6 5LR, Berkshire, UK.

SUMMARY

This review describes why polyphenols are important in animal nutrition. It also discusses the latest ideas on the reasons for their synthesis in plants. These ideas question their suggested role as chemical agents produced by plants to defend themselves against herbivores', and attribute their synthesis to stress conditions (e.g. nutrient imbalances during growth, high temperatures and light intensity, herbivore attack). Reference is made to the implications for breeding programmes which aim to improve the nutritive value of crop residues.

Some detailed physico-chemical studies of the interactions between polyphenols, carbohydrates and proteins are described. The following structural features enhance these molecular interactions: increasing molecular size, conformational flexibility, and low water solubility. Some carbohydrates are thought to form inclusion complexes with phenols and this capacity enhances the molecular interaction. Other factors important in the binding are the relative concentrations of polyphenols, carbohydrates and proteins, pH value, presence of metal ions and surfactants.

Since polyphenols are fairly reactive compounds it is crucial that precautions are taken during their analysis. Therefore, appropriate extraction, separation and analytical procedures are summarised and references are made to plant species in which polyphenols have so far been found. The list of new compounds and even new types of polyphenols is constantly being expanded.

This review finally describes some chemical treatments which have been applied to polyphenol-rich feeds and concludes by offering suggestions for further investigation.

INTRODUCTION

The efficient utilization of cereal crop residues and industrial byproducts is important in both developing and developed countries. In developing countries animal feeds are scarce and in developed countries burning of straw has become an environmentally unacceptable means of disposal. Crop residues are potentially rich sources of energy because up to 80% of their dry matter consists of polysaccharides; and some industrial byproducts contain high levels of protein (e.g. sal seed meal, neem cake). However, the digestion of these polysaccharides and proteins can be adversely affected by the presence of phenolic compounds.

There are several different types of phenolic compounds which have been implicated in both anti- and pro-nutritional responses: low molecular weight phenolic acids, aldehydes, and the so-called condensed and hydrolysable tannins.

There is no exact definition of the term "tannin", but we know that tannins contain phenolic groups. However, not all phenols are tannins! The original meaning is a compound able to convert hide to leather. This definition was subsequently modified to "a compound that binds proteins".[1] Recent detailed studies have shown that protein binding is at least in part a function of the chemical structures of both the phenol and protein and, in addition, it is dependent on other solution parameters.[2,3] The word "tannin" will therefore be avoided here except when referring to condensed (*syn.* proanthocyanidins) or hydrolysable tannins (*syn.* esters of gallic or ellagic acid). It is important that we have a thorough understanding of the basic binding mechanisms so that crop residues with improved nutritional characteristics can be produced.

IMPORTANCE

Nutritional effects

Several feeds have been examined which contain appreciable amounts of polyphenols: sainfoin,[4,5] *Lotus* spp.[6] and *Acacia seyal* feeds[7] have enhanced nitrogen retention in ruminants compared with similar feeds low in polyphenols. This has been attributed to an increased supply of amino acids to the small intestine as a result of binding of polyphenols to plant proteins, thus protecting the proteins from proteolysis in the rumen.[8,9] Thus, some polyphenols are important in the prevention of bloat.[10] However, other feeds containing high levels of polyphenols have lowered N-retention within ruminants.[7,11,12]

The digestion of carbohydrates can also be affected by polyphenols.[6,13,14] It would appear that polyphenols became irreversibly bound to polysaccharides inhibiting their subsequent fermentation in the rumen and lower intestine. This is in contrast to polyphenol-protein complexes, some of which dissociate in the hindgut, e.g. sainfoin polyphenol-protein complexes.[8]

It is thought that polyphenols may affect carbohydrate or protein digestion in several ways. Apart from complexing the substrates, polyphenols can interact with salivary proteins,[15,16] digestive enzymes or microorganisms.[11,16-19] Polyphenols may also decrease palatability[20,21] as they produce a bitter or astringent taste sensation.[22] They may therefore be responsible for lower intakes.[23] The digestion of feeds may also be impaired by physical effects which have been attributed to polyphenols: "woodiness" in plants has been related to the presence of condensed tannins.[24] Similarly, bird resistant sorghum grains, which have high levels of condensed tannins, possess a special testa layer.[25]

Although some attempts have been made at elucidating the structure-activity relationships of polyphenols,[26,27] no clear trends have yet been established.

Role of polyphenols in plants - implications for breeding programmes

It has been suggested that polyphenolic compounds are synthesised by plants 'in order to defend themselves against herbivore attack,[28,29] or against microbial attack[30] through their astringent taste and by interfering with the digestive enzymes of predators. However, several workers found no correlation between phenolic contents and herbivore attack.[31,32]

The principal biosynthetic thrust in plants is towards the production of higher molecular weight polyphenols which are the metabolic end-products.[33] However the ultimate purpose of these end-products is not yet known. Based on the above defence hypothesis one might assume therefore that such end-products are produced by plants because they are effective compounds in precipitating proteins of predators. The same authors[33] investigated the binding strength between various phenolic precursors and end-products with the protein bovine serum albumin (BSA). They found that one common phenolic precursor, pentagalloyl glucose, bound to the protein much more strongly than most end-products. This result casts doubt on the defence hypothesis.

A subsequent study[34] re-examined the evidence for the defence hypothesis and arrived at an alternative explanation for the apparently negative correlation between polyphenolic contents and predation by insects. The authors concluded that large quantities of non-extractable condensed tannins are formed during leaf development and that these contribute to leaf toughness, thus lowering the degree of insect attack.

Leaf toughness has been identified by some workers as the most important factor in herbivore attack, followed by fibre content.[31,35] As mentioned earlier, it was observed[24] that these polyphenolic compounds are often associated with woodiness in plants. Haslam and co-workers now draw attention to the possibility that condensed tannins which are covalently attached to polysaccharides may in fact contribute to lignification in plant tissues.[34]

Having questioned the "defence hypothyesis" an alternative explanation of the role of polyphenols in plants is needed. Haslam[36] reviewed the relationship between secondary and primary metabolism. The available evidence strongly suggests that secondary metabolism serves to maintain primary metabolism in circumstances not propitious for growth.[136] Nutrients stress, e.g. N, P, K, S-deficiencies, usually results in greater concentrations of phenolic compounds.[37,38] However, lack of water produces no definite trend on phenolic concentrations.[38] High light intensity, high temperature and severe drought were also linked to high phenolic contents.[39] The digestibility of fibre is limited by the extent of lignification, which is in turn greatly affected by environmental conditions such as climate and management.[40] Increased lignification of plant cell walls is apparently an effect of higher environmental temperatures possibly acting on plant enzymes.[40] The plant's ability to shunt primary metabolites into secondary metabolites may therefore be an adaptive mechanism which becomes important under stress conditions such as poor growing conditions or herbivore attack.

This hypothesis has important implications for breeding programmes which are explained below. The synthesis of polyphenolic compounds (e.g. monomeric flavonoids) is under enzymatic control which is genetically governed[41] and the capacity to synthesise higher molecular weight polyphenols is also a genetically inherited property as shown in the case of sorghum grain.[25] Even the degree of polymerisation may be genetically inherited as indicated by the existence of group II and group III sorghums.[42]

During biosynthesis, low molecular weight precursors of condensed and hydrolysable tannins are produced initially followed by higher molecular weight polyphenols at later stages of plant development.[43] Although many enzymes mediating the synthesis of secondary metabolites have very broad specificity,[36] there are other enzymes which are extremely substrate specific.[41] There is a clear need to understand better the effects of stress on the biosynthesis of polyphenols. In particular, which phenolic compounds increase under stress? Do stress conditions result in the accumulation of low molecular weight precursor or end-products? If the objective is to lower total polyphenolic concentrations, breeding programmes may be able to exploit genetic variability between plants causing accumulation of precursor compounds. On the other hand, if the chemical nature of the end-products needs alteration in order to achieve more beneficial nutritional effects, then the genetic variability within plant families needs to be explored. Such variability has, for example, been found in sorghum,[42] in cinnamon trees[44] and in oak.[34] It is likely, that changes arising from breeding crop residues with improved nutritional qualities need not necessarily result in lower grain yields.

Genetic engineering techniques have also been used; a colour change was created in petunia flowers.[41] Genes encoding a dihydroflavonol reductase were transferred from maize into petunia, thus enabling the synthesis of pelargonidin 3-glucoside from dihydrokaempferol. The expression of genes governing condensed tannin synthesis can occur in some but not other parts of plants. Clover, for example, synthesises condensed tannins in petals and stems but not in leaves, where they might be desirable to prevent bloat.[10] Similarly, high molecular weight flavan-3-ols are found in sorghum grain but not in sorghum leaves.[45] More work is needed on the mechanism of flavonoid gene expression. Light has been identified as one factor responsible for the expression of flavonoid synthesis.[46] Apparently, the degree of cell differentiation may also be important. Plant cells in tissue culture can produce secondary metabolites which differ from those produced by normal plants. Hydrolysable tannin production is inhibited and only a small quantity of precursor compound is synthesised in tissue cultures of *Quercus robur*.[36] However, tissue cultures of sorghum callus synthesised condensed tannins[47] which apparently do not occur in sorghum leaves.[45] It has been suggested that less differentiated cells and morphologically undeveloped callus tissues recognise and respond to stress differently than normal plants by producing condensed tannins.[47]

INTERACTIONS BETWEEN PHENOLIC COMPOUNDS AND CELL CONSTITUENTS

Recently, detailed chemical investigations of interactions between polyphenols, proteins and carbohydrates have been made and these will be described next. The binding mechanisms need investigating in order to explain why some polyphenols have pro-nutritional effects (e.g. sainfoin) and others have anti-nutritional effects (e.g. sorghum).

Carbohydrates

It would appear that not only low molecular weight[48] but also higher molecular weight polyphenols are covalently attached to carbohydrates. Some highly purified preparations of condensed tannins consistently contain sugar residues.[49,50] However, it was not possible to determine the linkages because of a similar lability of the glucosidic and inter-flavanol bonds. Available evidence also suggests that some condensed tannins are covalently bound to cell wall polysaccharides.[47,51,52] Both flavan 3-O-[49] and 4-O-linkages[51] to carbohydrates have been proposed. Novel flavanol C-glucosides were easily extracted with water from cinnamon bark and the carbon-carbon linkages were identified as 8-C and 6-C. These compounds were also synthesised under acid conditions from (-)epicatechin and glucose.[53]

Difficulties experienced during the extraction of polyphenols from plant materials may also be caused by strong non-covalent binding between polyphenols and polysaccharides.[54] It has been shown[55,56] that non-covalent binding to carbohydrates is enhanced by the following polyphenol characteristics: increasing molecular size, conformational flexibility, low water solubility and the capacity to develop a secondary structure which allows the formation of inclusion complexes. Increased adsorption was also found at higher pH values, i.e. a slight increase at pH 4.0 compared to pH 2.2[55] and marked increase at pH 13 compared to pH 7.[54] These results may help to explain why starches from different sources bind polyphenols to differing degrees.[13,54,57] However, no studies have yet been made on the digestibility of carbohydrate in these sort of complexes - which often are soluble - where phenols are included into cavities compared to complexes where the phenols are adsorbed onto the surfaces (e.g. with cellulose).[58]

Proteins

The interaction between polyphenols and proteins has been recognised for much longer than that between polyphenols with polysaccharides. However, we do not yet understand the differences in digestibility caused by feed containing similar amounts of polyphenols. As mentioned earlier, polyphenols have both beneficial and detrimental effects on nitrogen metabolism in ruminants.

The complexes formed by the interaction of proteins and phenols in solution can be either soluble or insoluble.[59,60] However, insoluble (non-extractable) polyphenols can also complex soluble proteins[61] and *vice versa*.[62] Furthermore, some water insoluble hydrolysable tannins are kept in solution by amino acids.[63] It has been tentatively suggested that the interactions are not the same in soluble and insoluble complexes and that the binding strength may be stronger within insoluble complexes.[60]

The effect of complex formation and the degree of solubility of these complexes on the digestibility of proteins and on the activity of enzymes is not yet known. It has been suggested[64] that soluble complexes may be important in overcoming the anti-nutritional effects of some polyphenols. Complexation of enzymes by polyphenols may inhibit,[65] reduce[54,66] or even enhance enzyme activity.[60,67]

Complex formation depends on both the polyphenol and the protein concentrations resulting in variable stoichiometries (e.g. ranging between 1:60 and 1:120 for the protein:polyphenol ratio).[2,60] Precipitation is thought to occur when a hydrophobic outer layer is formed. Thus at appropriate concentrations even simple phenols, such as pyrogallol and resorcinol, can precipitate proteins from solution[2]. This example illustrates best why the term tannin is obsolete.

pH is an important solution parameter which affects the extent of complexation and thus the digestion of bound substrates.[8,68] Precipitation is greatest at a pH-value which is within one unit of the isoelectric point of the protein.[69,70,71] Hydrogen bonding and hydrophobic interaction have been implicated in complex formation by several workers.[72] Characteristically, hydrogen bonding depends on pH, whereas hydrophobic bonding is much less dependent on pH.[73] Hydrophobic bonds in polyphenol-protein complexes have been disrupted by surfactant.[74] Polyphenol-protein complexes are also affected by pH. These findings suggest that both bonding mechanisms are important during complexation. Their relative importance has not been resolved, but thermodynamic measurements in the case of bovine serum albumin and hydrolysable tannins indicate that hydrophobic interaction is not predominant.[58] X-ray analysis of crystalline complexes of polyphenol and protein analogues[75,76] and NMR studies of aqueous solutions[56] clearly demonstrated the existence of weak hydrophobic interaction between aryl rings which were stacked above each other. In addition, there was hydrogen bonding, and when K^+ ions were present these were surrounded by 7 coordinated oxygen atoms.[76] This type of stacking arrangement also occurs when polyphenols self-associate.[77] Alkali metals promote polyphenol-protein complexation[3] and self-association,[76] whereas urea and dimethylsulfoxide disrupt these aggregates.[77]

Hagerman and Klucher[73] suggested that condensed and hydrolysable tannins have a similar interaction mechanism with proteins. Some degree of confusion exists over the relative importance of polyphenol or protein structure to the binding strength in complexes. Some work has shown that interactions are highly protein-specific and not polyphenol-specific,[73] whilst others found that the relative ranking order in binding strengths of polyphenols was little affected by the type of protein involved.[58] However, studies using a mixture of different polyphenols and proteins clearly indicate that complex formation is specific for both polyphenols and proteins.[78]

As polyphenols interact with proteins as well as polysaccharides, it is of no surprise that the presence of polysaccharides modified the interaction between polyphenols and proteins.[64] The authors suggested that oligosaccharides in glycoproteins enhance the affinity and selectivity of proteins for polyphenols and at the same time increase the solubility of such complexes. On the other hand, Strumeyer and Malin[79] suggested that the high content of carbohydrates in two glycoproteins, yeast invertase and tannase, were responsible for their resistance to denaturation by polyphenols. Modifying effects are especially noticeable where polysaccharides can develop a secondary structure containing hydrophobic cavities which lead to inclusion complexes[56,57] rather than surface adsorption, as is the case with cellulose.[58]

Analogous with carbohydrates, molecular size and conformational flexibility of the polyphenol have major effects on the strength of protein binding.[33,55] Proteins with open, loose conformations interact much more strongly with sorghum polyphenols than globular proteins.[64] The great specificity of some polyphenols for certain proteins[32,64,65] may be likened to the specificity between enzymes and their substrates. Consequently, some proteins are preferentially precipitated from solution even in the presence of excess amounts of other proteins.[65]

One may speculate that the binding strength in such complexes has important implications on the degradability of bound substrates by enzymes of micro-organisms.[68] This hypothesis needs testing. Denaturing of proteins by tannins has been suggested as the possible cause for enhanced digestion of some complexes.[60]

Studies are also needed to assess any reactions within the polyphenol-protein complex that may occur during digestion. Beart et al.[51] predicted that covalent bonds could be formed in such complexes at the pH values prevailing in digestive tracts of ruminants. However, this proposition might contradict the pro-nutritional effects observed with sainfoin.[10]

EXTRACTION AND SEPARATION OF POLYPHENOLS

Polyphenols are polar compounds and therefore extractable with water, methanol or acetone.[53,80-84] However, methanolic HCl is required to extract condensed tannins from certain sorghum grains.[84] Ethyl acetate extracts low molecular weight condensed tannins (monomers, dimers) from aqueous solutions[82], whereas partitioning between n-butanol and water afforded trimeric to pentameric proanthocyanidins.[44] Aqueous acetone is especially effective in dissolving polyphenols from some plants;[10,85] it extracted condensed tannin with molecular weights up to 28,000 daltons corresponding to 80 flavan-3-ol units.[10]

It is important to take precautions when extracting and storing polyphenols[11,86] as they are quite sensitive to UV light and oxygen. Acetone is a good solvent as oxidation is minimised.[87] Hydrolysable tannins may even react with methanol at room temperature and pH 6.[88]

Preliminary purification of polyphenols is routinely performed on Sephadex LH-20[81,83] by elution with methanol and aqueous acetone. The crude fractions can be further purified by TLC[82] or gel permeation chromatography.[44,89,90] Finally, high performance liquid chromatography (HPLC) may be used semi-preparatively[91,92] in difficult separations for the isolation of pure compounds.

There are several recent reviews[93,94] which describe applications of HPLC to flavonoid analysis. Separation of short chain proanthocyanidins could easily be achieved by HPLC, but larger polymers were less well resolved and complex peaks with complex elution patterns were obtained.[17,95] Co-chromatography with known compounds in combination with wavelength ratioing is suitable for checking purities and identities of the isolated compounds.[96]

Due to the high resolution power of HPLC some pure compounds (hydrolysable tannins) can give multiple peaks in chromatograms which may arise from different stereoisomers.[97] Multiple peaks in chromatograms of catechin gallates were attributed to migration of the depsidically linked gallic acid residues under the acidic conditions of the eluting solvent.[91]

ANALYSIS AND IDENTIFICATION OF POLYPHENOLS

Quantitative estimates of polyphenols have been attempted in order to assess their nutritional effects. However, the literature on the effects of phenolic compounds in animal nutrition has many examples of the inappropriate use of analytical methods resulting in unsupported conclusions. This often stems from a poor understanding of the reactions between phenolic compounds and their detection reagents, i.e. phenolic compounds have been estimated using non-specific methods. The quantification of polyphenols with great structural diversities is problematic because of a general lack of reference compounds. In various assays, gallic acid,[98] catechin[99] and chlorogenic acid[100] have been used as reference standards, although they bear little structural similarity to the polyphenols being analysed. Tannic acid[11] is also a poor standard because it tends to be a mixture of different compounds, the relative proportions of which vary between samples.[101] Using catechin as a standard in the vanillin test over-estimates the concentration of condensed tannins because the reaction kinetics are different for the monomer and the oligomers.[99] A new procedure for precipitating all soluble phenols from plant extracts with ytterbium acetate has been developed.[102] This method does not require a reference compound as the phenols are quantified directly by gravimetry. Precipitation was complete when phenols accounted for more than 16% of the dry matter. The precipitates contained flavan-3-ols, condensed tannins, gallic acid and hydrolysable tannins, catechin gallates, flavonols and their glycosides (Mueller-Harvey, Reed and

Hartley, unpublished data). Little protein or chlorophyll appeared to be co-precipitated with the phenols. Furthermore, it is likely that different phenolic compounds may have different nutritional effects and therefore the quantification of "total phenols" may not be what is required. This make comparative assessments of different feeds for their nutritional response extremely difficult.

HPLC is an excellent tool for quantitative measurements if reference compounds are available. Quantification of condensed and hydrolysable tannins, however, is hampered by the lack of such standards. In addition, several discrepent UV-absorbance values (280 nm) have been reported for pure procyanidins.[83,103,104,105] The trend is towards lower and lower E-values which may reflect the attainment of better and better purities or, more likely, it may be due to the age of the polymers. E-values at 550 nm obtained after n-butanol/HCl treatment, depended on the age of the proanthocyanidins.[106] It is interesting to note, that in one detailed study[105] E-values were the same for the monomer up to the tetramer on a weight basis. This suggests that it might be possible to quantify even broad peaks in HPLC chromatograms by peak area provided the homogeneity of the proanthocyanidins is known. It has also been shown[83] that UV absorbance of polymers containing a range of procyanidins and prodelphinidins are additive as their chromophores behave like uncoupled oscillators. The monomeric composition of polymers may also be estimated by ^{13}C NMR[83] and n-butanol/Fe^{3+}/HCl hydrolysis.[106]

Although plant phenolic compounds have been measured colorimetrically using the Folin-Denis reagent,[107] it must be remembered that this reagent reacts non-stoichiometrically with phenolic and other OH groups and that several reducing agents interfere.[98] Despite its lack of specificity for polyphenols, this reagent has often been used to measure "tannin" content in forage crops because of its ease of use.[108] A modified Folin reagent[109] gives a better estimate of total phenolic groups[108] because it gives a greater colour response with phenols and a lesser response to non-phenolic compounds. Another general detection reagent for phenolic groups is the Prussian blue reagent,[110] but it has the disadvantage of widely varying sensitivity for different compounds.

Vanillin reacts under acidic conditions with only one group of polyphenols the flavanols. It is specific for a narrow range of flavanols (including condensed tannins) and dihydrochalcones[107] but only when performed under the recommended conditions,[111] The widely used hydrolysis of proanthocyanidins in alcohol to yield coloured anthocyanidins has recently been re-examined and modified.[106] The reaction is now reproducible and the yield of anthocyanidin is related to the concentration of proanthocyanidins.

The quantification of hydrolysable tannins may be attempted by measuring the release, after hydrolysis, of gallic or ellagic acid and glucose or quinic acid using HPLC. Ellagitannins have been measured colorimetrically with nitrous acid.[112]

A completely different approach has been to measure the protein binding capacity of polyphenols.[69,113,114,115] In this procedure, many phenols with low molecular weights are less likely to be measured. The limitations of this approach are obvious from the above discussion of phenol-protein binding which is dependent on solution conditions and also on the structures of the phenols and proteins. For example, the choice of hide powder[113] has been criticised because it selectively binds only some compounds from a mixture of tannic acids. In addition, binding of polyphenols to one particular protein may not be related to its effects *in vivo* on salivary proteins, digestive enzymes or rumen micro-organisms.

A consecutive analysis approach for the preliminary characterisation of polyphenols is based on R_f-values and various spray reagents applied to TLC.[116,117] Isolated compounds are subjected to so-called shift reagents and their UV spectra are recorded.[81,118,119] The shifts obtained for the UV spectral bands are used to determine the oxidation state of the flavonoid C ring and the positions of phenolic groups, carbohydrates or methoxy groups. Structure determinations are completed by proton and carbon nuclear magnetic resonance[118] or mass spectrometry, especially in the fast atom bombardment mode.[91] The large molecular weights of some polyphenols have also been measured using ^{13}C NMR.[50,83]

OCCURRENCE OF POLYPHENOLS

The occurrence of condensed and hydrolysable tannins has been summarised for a wide range of different plant species.[80,120] However, new compounds are still being identified. Amongst the Gramineae, condensed tannins have been found in sorghum grain.[121,122] Although sorghum leaves are extremely rich in polyphenols, no procyanidins could be detected.[45] When the neutral detergent fibre fraction of sorghum crop residues was subjected to hydrolysis in butanol/HCl compounds absorbing at 550 nm were formed which indicated that proanthocyanidins of some sort were present.[52] Polyphenols in general and proanthocyanidins in particular were higher in bird resistant than in non-bird resistant sorghum varieties.[52] Proanthocyanidins also occur in millet[123] and barley grain.[124] It would appear that crop residues of these species have not yet been examined for proanthocyanidins.

Condensed tannins are quite important in several pasture legumes.[10,125,126,127] As many industrial by-products are produced in large quantities by the edible oil industry, these have been used as animal feeds, but their high polyphenol levels limited the use of peanut skins[90] and sal seed cake.[128]

Soft fruits and tree leaves - often used by browsing animals - are rich sources of both condensed and hydrolysable tannins and of an intermediate group, the flavan-3-ol gallates.[17,49,50,53,129-133]

TREATMENTS OF FEEDS RICH IN POLYPHENOLS

Crop residues and many industrial by-products could be utilised as anima feeds if the anti-nutritional effects of polyphenols could be overcome. It ma be possible to breed for varieties with lower polyphenol contents or wit different chemical structures of the polyphenols. Treating the product chemically may present another possible avenue. Such treatments must b cheap and easy to use, if they are to be applied at the farm level.

Most efforts so far have concentrated on finding treatments for sorghur grain polyphenols, and only these will be summarised here. Addition of ure to sorghum grain enhanced the *in vitro* digestibility of dry matter; however, th *in vitro* protein digestibility remained unaffected.[134] Improved chick growt was obtained after adding methionine or polyvinylpyrrolidone to sorghur grain.[135]

Formaldehyde treatment lowered the amount of protein being precipitate by polyphenols[136] but no digestibility trials were conducted to assess th nutritional effects of the treated samples. Ensiling,[137] anaerobic storage unde CO_2 after adding water, HCl or NaOH,[138] and heating at 70°C and pH 6[139] enhanc the *in vitro* dry matter and protein digestibility. It should be pointed out her that treatments which lower extractable polyphenol levels do not always resu in improved digestibilities.[140] However, soaking with 0.1M NH_3 in water fc several hours at room temperature greatly improved the nutritional quality c sorghum grain.[140]

As the astringent taste of polyphenols in unripe fruits may be masked maturity by soluble pectins,[57] it may be possible to mix feeds which are rich i polyphenols with industrial by-products which contain soluble oligomer carbohydrates. The partial hydrolysis of polysaccharides, for example, in cro residues could be an alternative means of bringing about phenol-carbohydra complexation. The resulting phenol-carbohydrate complexes will have to t examined for their nutritional effects and digestibility.

SUGGESTIONS FOR FURTHER RESEARCH

Despite the fact that many nutritional trials with polyphenol-rich feeds hav been conducted, that many polyphenols have been identified chemically an that many proteins have *in vitro* been precipitated with polyphenols, a basi understanding is still lacking of which phenols produce pro-nutritional an which phenols produce anti-nutritional effects. We need to know ho interactions with polyphenols affect the digestibility of feed proteins an carbohydrates, i.e. to what extent is the digestibility affected by the bindir strength within complexes and the solubility of the complexes. So far, only limited amount of work has been done on the action of polyphenols on rume microorganisms and on digestive enzymes. This needs further study.

Most reports of the links between nutrient stress and high polyphenol concentrations stem from field observations. Plant growth experiments under controlled conditions are therefore needed in order to assess what forms of stress enhance polyphenol synthesis and for what reasons. These should be done with the aim of designing good management practices that result in lower anti-nutritional polyphenol levels in plants.

It is likely that in the not too distant future crop residues with improved nutritive value can be obtained through breeding experiments and genetic engineering.

Acknowledgements

I would like to thank Dr. A.B. McAllan for his succinct comments on earlier draft versions and Mr. P.M.S. Blackwell for his help with the literature search.

REFERENCES

1. Swain, T. and Bate-Smith, E.C. (1962). Flavonoid compounds. *Comprehensive Biochemistry,* **3**, 755-809.

2. McManus, J.P.; Davis, K.G.; Lilley, T.H. and Haslam, E. (1981). The association of proteins with polyphenols. *Journal of the Chemical Society, Chemical Communications* 309-311.

3. Martin, M.M.; Rockholm, D.C. and Martin, J.S. (1985). Effects of surfactants, pH and certain cations on precipitation of proteins by tannins. *Journal of Chemical Ecology* **11,** 485-494.

4. Thomson, D.J.; Beever, D.E.; Harrison, D.G.; Hill, I.W. and Osbourn, D.F. (1971). The digestion of dried lucerne and sainfoin by sheep. *Proceedings of the Nutrition Society* **3**, 14A.

5. Egan, A.R. and Ulyatt, M.J. (1980). Quantitative digestion of fresh herbage by sheep. VI. Utilisation of nitrogen in five herbages. *Journal of Agricultural Science (Cambridge)* **94**, 47-56.

6. Barry, T.N.; Manley, T.R. and Duncan, S.J. (1986). The role of condensed tannins in the nutritional value of *Lotus pedunculatus* for sheep. 4. Sites of carbohydrate and protein digestion as influenced by dietary reactive tannin concentration. *British Journal of Nutrition* **55,** 123-137.

7. Reed, J.D. and Soller, H. (1987). Phenolics and nitrogen utilization in sheep fed browse. In *Herbivore nutrition research* M. Rose (ed.). 2nd International Symposium (Queensland), July 1987. pp. 47-48.

8. Jones, W.T. and Mangan, J.L. (1977). Complexes of the condensed tannins of sainfoin (*Onobrychis viciifolia* Scop.) with Fraction 1 leaf protein and with submaxillary mucoprotein, and their reversal by polyethylene glycol and pH. *Journal of the Science of Food and Agriculture* **28,** 126-136.

9. Reid, C.S.W.; Ulyatt, M.J. and Wilson, M.J. (1974). Plant tannins, blo and nutritive value. *Proceedings of the New Zealand Society of Anim Production* **34,** 82-93.

10. Jones, W.T.; Broadhurst, R.B. and Lyttleton, J.W. (1976). The condense tannins of pasture legume species. *Phytochemistry* **15,** 1407-1409.

11. McLeod, M.N. (1974). Plant tannins - their role in forage quality. *Nutritic Abstracts and Reviews* **44,** 803-815.

12. Kuman, R. and Singh, M. (1984). Tannins: their adverse role in rumina nutrition. *Journal of Agricultural and Food Chemistry* **32,** 447-453.

13. Deshpande, S.S. and Salunkhe, D.K. (1982). Interactions of tannic ac and catechin with legume starches. *Journal of Food Science* **4** 2080-2083.

14. Barry, T.N. and Manley, T.R. (1986). Interrelationships between th concentrations of total condensed tannins, free condensed tannins ar lignin in *Lotus* sp. and their possible consequences in ruminant nutritio *Journal of the Science of Food and Agriculture* **37,** 248-254.

15. Mehansho, H.; Hagerman, A.; Clements, S.; Butler, L.; Rogler, J. ar Carlson, D. (1983). Modulation of proline-rich protein biosynthesis in r parotid glands by sorghums with high tannin levels. *Proceedings Nation Academy of Sciences (USA)* **80,** 3948-3852.

16. Burritt, E.A.; Malechek, J.C. and Provenza, F.D. (1987). Changes concentrations of tannins, total phenolics, crude protein, and *in vitr digestibility of browse due to mastication and insalivation by cattl *Journal of Range Management* **40,** 409-411.

17. Mueller-Harvey, I.; McAllan, A.B.; Theodorou, M.K. and Beever, D.l (1988). Phenolics in fibrous crop residues and plants and their effects c the digestion and utilisation of carbohydrates and proteins in ruminant In *Plant Breeding and the Nutritive value of Crop Residues.* J.D. Ree B.S. Capper and P.J.H. Neate (eds.). International Livestock Centre f Africa. Addis Ababa, Ethiopia pp. 97-132.

18. Grant, W.D. and McMurtry, C.B. (1978). Effects of condensed tannir on the growth of microorganisms. In *Microbial ecology.* M.W. Lou and J.A.R. Miles (eds.). Springer Verlag, Berlin. pp. 427-430.

19. Mori, A.; Nishino, C.; Enoki, N. and Tawata, S. (1987). Antibacteri activity and mode of action of plant flavonoids against *Proteus vulgar* and *Staphylococcus aureus. Phytochemistry* **26,** 2231-2234.

20. Provenza, F.D. and Malechek, J.C. (1984). Diet selection by domest goats in relation to blackbrush twig chemistry. *Journal of Applied Ecolo* **21,** 831-841.

21. Cooper, S.M. and Owen-Smith, N. (1985). Condensed tannins defer feeding by browsing ruminants in a South African Savanna. *Oecologia* **67,** 142-146.

22. Lea, A.G.H. (1984). Tannin and colour in English cider apples. *Fluessiges Obst* **8,** 1-5.

23. Barry, T.N. and Duncan, S.J. (1984). The role of condensed tannins in the nutritional value of *Lotus pedunculatus* for sheep. 1. Voluntary intake. *British Journal of Nutrition* **51,** 485-491.

24. Bate-Smith, E.C. and Lerner, N.H. (1954). 2. Systematic distribution of leucoanthocyanins in leaves. *Biochemistry Journal* **58,** 126-132.

25. Hahn, D.H.; Rooney, L.W. and Earp, C.F. (1984). Tannins and phenols of sorghum. *Cereal Foods World,* **29,** 776-779.

26. Cody, V.; Middleton Jr., E. and Harborne, J.B. (1986). *Plant Flavonoids in Biology and Medicine: Biochemical, Pharmacological and Structure-Activity Relationships.* Alan R. Liss, Inc., New York.

27. Williams, E.A. and Menary, R.C. (1988). Polyphenolic inhibitors of alpha-acid oxidase activity. *Phytochemistry* **27,** 35-39.

28. Feeny, P. (1976). Plant apparency and chemical defense. In: "Biochemical interactions between plants and insects" (J.W. Wallace and R.L. Mansell, eds.). Plenum, New York. *Recent Advances in Phytochemistry* **10,** 1-40.

29. Waterman, P.G. (1986). A phytochemist in the African rain forest. *Phytochemistry* **25,** 3-17.

30. Zucker, W.V. (1983). Tannins: Does structure determine function? An ecological perspective. *American Naturalist* **121,** 335-365.

31. Coley, D. (1983). Herbivory and defensive characteristics of tree species in a lowland tropical forest. *Ecological Monographs* **53,** 209-233.

32. Becker, P. and Martin, J.S. (1982). Protein-precipitating capacity of tannins in *Shorea (Dipterocarpaceae)* seedling leaves. *Journal of Chemical Ecology* **8,** 1353-1367.

33. Beart, J.E.; Lilley, T.H. and Haslam, E. (1985). Plant polyphenols - secondary metabolism and chemical defence: some observations. *Phytochemistry* **24,** 33-38.

34. Scalbert, A. and Haslam, E. (1987). Polyphenols and chemical defence of the leave of *Quercus robur. Phytochemistry* **26,** 3191-3195.

35. Feeny, P.P. (1970). Seasonal changes in oak leaf tannins and nutrients as a cause of spring feeding by winter moth caterpillars. *Ecology* **51,** 565-581.

36. Haslam, E. (1986). Secondary metabolism - fact and fiction. *Natural Product Reports* **3,** 217-249.

37. Barry, T.N. and Forss, D.A. (1983). The condensed tannin content of vegetative *Lotus pedunculatus,* its regulation by fertilizer application, and effect upon protein solubility. *Journal of the Science of Food and Agriculture* **34,** 1047-1056.

38. Gershenzon, J. (1983). Changes in the levels of plant secondary metabolites under water and nutrient stress. Ch.10 in: "Phytochemical adaptations to stress". Plenum Press, New York and London. *Recent Advances in Phytochemistry* **18,** 273-320.

39. Burns, R.E. (1966). Tannins in *Sericea lespedeza.* Georgia Agricultural Experiment Stations. Bulletin N.S. 164.

40. Van Soest, P.J. (1988). Effect of environment and quality of fibre on the nutritive value or crop residues. In *Plant Breeding and the Nutritive Value of Crop Residues.* International Livestock Center for Africa, Addis Ababa, Ethiopia pp 71-96.

41. Meyer, P.; Heidmann, I.; Forkmann, G. and Saedler, H. (1987). A new petunia flower colour generated by transformation of a mutant with a maize gene. *Nature* **330,** 667-678.

42. Bullard, R.W.; York, J.O. and Kilburn, S.R. (1981). Polyphenolic changes in ripening bird-resistant sorghums. *Journal of Agricultural and Food Chemistry* **29,** 973-981.

43. Swain, T. (1986). The evolution of flavonoids. In *Plant Flavonoids in Biology and Medicine: Biochemical, Pharmacological, and Structure-activity Relationships.* V. Cody, E. Middleton Jr. and J.B. Harborne (eds.). Alan R. Liss, Inc., New York. pp. 1-14.

44. Nonaka, R.; Morimoto, S. and Nishioka, I. (1983). Tannins and related compounds. Part 13. Isolation and structures of trimeric, tetrameric and pentameric proanthocyanidins from cinnamon. *Journal of the Chemical Society, Perkin Transations* **I,** 2139-2145.

45. Watterson, J.J. and Butler, L.G. (1983). Occurrence of an unusual leucoanthocyanidin and absence of proanthocyanidins in sorghum leaves. *Journal of Agricultural and Food Chemistry* **31,** 41-45.

46. Bolwell, G.P. (1988). Synthesis of cell wall components: aspects of control. *Phytochemistry* **27,** 1235-1253.

47. Oberthur, E.E.; Nicholson, R.L. and Butler, L.G. (1983). Presence of polyphenolic materials, including condensed tannins, in sorghum callus. *Journal of Agricultural and Food Chemistry,* **31,** 660-662.

48. Hartley, R.D.; Whatley, F.R. and Harris, P.J. (1988). 4,4'-Dihydroxytruxillic acid as a component of cell walls of *Lolium multiflorum. Phytochemistry* **27,** 349-351.

49. Porter, L.J.; Foo, L.Y. and Furneaux, R.H. (1985). Isolation of three naturally occurring 0-β-glucopyranosides of procyanidin polymers. *Phytochemistry* **24**, 567-569.

50. Shen, Z.; Haslam, E. and (in part) Falshaw, C. P. and Begley, M.J. (1986). Procyanidins and polyphenols of *Larix gmelini* bark. *Phytochemistry* **25**, 2629-2635.

51. Beart, J.E.; Lilley, T.H. and Haslam, E. (1985). Polyphenol interactions. Part 2. Covalent binding of procyanidins to proteins during acid-catalysed decomposition; observations on some polymeric proanthocyanidins. *Journal of the Chemical Society, Perkin Transactions* **II**, 1439-1443.

52. Reed, J.D.; Tedla, A. and Kebede, Y. (1987). Phenolics, fibre and fibre digestibility in the crop residue from bird resistant and non-bird resistant sorghum varieties. *Journal of the Science of Food and Agriculture* **39**, 113-121.

53. Morimoto, S.; Nonaka, G.-I. and Nishioka, I. (1986). Tannins and related compounds. XXXVIII. Isolation and characterization of flavan-3-ol glucosides and procyanidin oligomers from Cassia bark (*Cinnamomum cassia* Blume). *Chemical Pharmaceutical Bulletin (Tokyo)* **34**, 633-642.

54. Davis, A.B. and Hoseney, R.C. (1979). Grain sorghum condensed tannins. I. Isolation, estimation, and selective adsorption by starch. *Cereal Chemistry* **56**, 310-314.

55. McManus, J.P.; Davis, K.G.; Beart, J.E.; Gaffney, S.H.; Lilley, T.H. and Haslam, E. (1985). Polyphenol interactions. Part 1. Introduction: some observations on the reversible complexation of polyphenols with proteins and polysaccharides. *Journal of the Chemical Society, Perkin Transactions* **II**, 1429-1438.

56. Gaffney, S.H.; Martin R.; Lilley, T.H.; Haslam, E. and Magnolato, D. (1986). The association of polyphenols with caffeine and α- and β-cyclodextrin in aqueous media. *Journal of the Chemical Society, Chemical Communications* 107-109.

57. Ozawa, T.; Lilley, T.H. and Haslam, E. (1987). Polyphenol interactions: Astringency and the loss of astringency in ripening fruit. *Phytochemistry* **26**, 2937-2942.

58. Haslam, E. and Lilley, T.H. (1986). Interactions of natural phenols with macromolecules. In *Plant Flavonoids in Biology and Medicine: Biochemical, Pharmacological and Structure-activity Relationships* V. Cody, E. Middleton Jr. and J.B. Harborne (eds.). Alan R. Liss, Inc. New York. pp. 53-65.

59. Van Buren, J.P. and Robinson, W.B. (1969). Formation of complexes between protein and tannic acid. *Journal of Agricultural and Food Chemistry* **17**, 772-777.

60. Mole, S. and Waterman, P.G. (1985). Stimulatory effects of tannins and cholic acid on tryptic hydrolysis of proteins: ecological implications. *Journal of Chemical Ecology* **11,** 1323-1332.

61. Bate-Smith, E.C. (1977). Astringent tannins of *Acer* species. *Phytochemistry* **16,** 1421-1426.

62. Verzele, M.; Delahaye, P. and Van Damme, F. (1986). Determination of the tanning capacity of tannic acids by high-performance liquid chromatography. *Journal of Chromatography* **362,** 363-374.

63. Okuda, T.; Yoshida, T.; Hatano, T.; Ikeda, Y.; Shingu, T. and Inoue, T. (1986). Constituents of *Geranium thunbergii* Sieb. *et* Zucc. XIII. Isolation of water-soluble tannins by centrifugal partition chromatography, and biomimetric synthesis of Elaeocarpusin. *Chemical and Pharmaceutical Bulletin (Tokyo)* **34,** 4075-4082.

64. Asquith, T.N.; Uhlig, J.; Mehansho, H.; Putman, L.; Carlson, D.M. and Butler, L.G. (1987). Binding of condensed tannins to salivary proline-rich glycoproteins: the role of carbohydrate. *Journal of Agricultural and Food Chemistry* **35,** 331-334.

65. Butler, L.G.; Riedl, D.J.; Lebryk, D.G. and Blytt, H.J. (1984). Interaction of proteins with sorghum tannin: mechanism, specificity and significance. *Journal of American Oil Chemists' Society* **61,** 916-920.

66. Goldstein, J.L. and Swain, T. (1965). The inhibition of enzymes by tannins. *Phytochemistry* **4,** 185-192.

67. Mole, S. and Waterman, P.G. (1987). Tannic acid and proteolytic enzymes. Enzyme inhibition or substrate deprivation? *Phytochemistry* **67,** 99-102.

68. Martin, J.S. and Martin, M.M. (1983). Tannin assays in ecological studies. Precipitation of ribulose-1,5-biphosphate carboxylase/oxygenase by tannic acid, quebracho, and oak foliage extracts. *Journal of Chemical Ecology* **9,** 285-294.

69. Hagerman, A.E. and Butler, L.G. (1978). Protein precipitation method for the quantitative determination of tannins. *Journal of Agricultural and Food Chemistry* **26,** 809-812.

70. Watanabe, T.; Mori, T.; Tosa, T. and Chibata, I. (1981a). Characteristics of immobiloised tannin for protein adsorption. *Journal of Chromatography* **207,** 13-20.

71. Watanabe, T.; Mori, T.; Tosa, T. and Chibata, I. (1981b). Adsorption specificity of immobilized tannin for proteins and other organic compounds. *Agricultural and Biological Chemistry* **45,** 1001-1003.

72. Artz, W.E.; Bishop, P.D.; Dunker, A.K.; Schanus, E.G. and Swanson, B.G. (1987). Interaction of synthetic proanthocyanidin dimer and trimer with bovine serum albumin and purified bean globulin Fraction G-1. *Journal of Agricultural and Food Chemistry* **35,** 417-421.

73. Hagerman, A.E. and Klucher, K.M. (1986). Tannin-protein interactions. In *Plant Flavonoids in Biology and Medicine: Biochemical, Pharmacological, and Structure-activity Relationships.* V. Cody, E. Middleton Jr. and J.B. Harborne (eds.). Alan R. Liss, Inc., New York. pp. 67-76.

74. Martin, M.M. and Martin, J.S. (1984). Surfactants: their role in preventing the precipitation of proteins by tannins in insect guts. *Oecologia (Berlin)* **61,** 342-345.

75. Martin, R.; Lilley, T.H.; Bailey, N.A.; Falshaw, C.P.; Haslam, E.; Magnolato, D. and Begley, M.J. (1986). Polyphenol-caffeine complexation. *Journal of the Chemical Society, Chemical Communications,* 105-106.

76. Martin, R.; Lilley, T.H.; Falshaw, C.P.; Haslam, E.; Begley, M.J. and Magnolato, D. (1987). The caffeine-potassium chlorogenate molecular complex. *Phytochemistry* **26,** 273-279.

77. Brouillard, R. (1983). The *in vitro* expression of anthocyanin colour in plants. *Phytochemistry* **22,** 1311-1323.

78. Asquith, T.N. and Butler, L.G. (1986). Interactions of condensed tannins with selected proteins. *Phytochemistry* **25,** 1591-1593.

79. Strumeyer, D.H. and Malin, M.J. (1970). Resistance of extracellular yeast invertase and other glycoproteins to denaturation by tannins. *Biochemical Journal* **118,** 899-900.

80. Thompson, R.S.; Jacques, D.; Haslam, E. and (in part) Tanner, R.J.N. (1972). Plant proanthocyanidins. Part 1. Introduction: the isolation, structure and distribution in nature of plant procyanidins. *Journal of the Chemical Society, Perkin* **I,** 1387-1399.

81. Strumeyer, D.H. and Malin, M.J. (1975). Condensed tannins in grain sorghum: isolation, fractionation and characterization. *Journal of Agricultural and Food Chemistry* **23,** 909-914.

82. Foo, L.Y. and Porter, L.J. (1978). Prodelphinidin polymers: Definition of structural units. *Journal of the Chemical Society, Perkin* **I,** 1186-1190.

83. Czochanska, Z.; Foo, L.Y.; Newman, R.H. and Porter, L.J. (1980). Polymeric proanthocyanidins. Stereochemistry, structural units and molecular weight. *Journal of the Chemical Society, Perkin* **I,** 2278-2286.

84. Asquith, T.N.; Izuno, C.C. and Butler, L,G. (1983). Characterization of the condensed tannin (proanthocyanidin) from a group II sorghum. *Journal of Agricultural and Food Chemistry* **31,** 1299-1303.

85. Bate-Smith, E.C. (1975). Phytochemistry of proanthocyanidins. *Phytochemistry* **14,** 1107-1113.

86. Gartlan, J.S.; McKey, D.B.; Waterman, P.G.; Mbi, C.N. and Struhsackers, T.T. (1980). A comparative study of the phytochemistry of two African rain forests. *Biochemical Systematics and Ecology* **8**, 401-422.

87. Schanderl, S.H. (1970). Tannins. In: 'Methods of food analysis' (M.A. Joslyn, ed.) 2nd edition. Academic Press, London. p.701.

88. Haslam, E. (1966). *Chemistry of Vegetable Tannins.* Academic Press, London and New York.

89. Derdelinckx, G. and Jerumanis, J. (1984). Separation of malt and hop proanthocyanidins on Fractogel TSK HW 40(S). *Journal of Chromatography* **285**, 231-234.

90. Karchesy, J.J. and Hemingway, R.W. (1986). Condensed tannins: (4β → 8; 2β → O → 7)-linked procyanidins in *Arachis hypogea* L. *Journal of Agricultural and Food Chemistry* **34**, 966-970.

91. Self, R.; Eagles, J.; Galletti, G.C.; Mueller-Harvey, I; Hartley, R.D.; Lea, A.G.H.; Magnolato, D.; Richli, U.; Gujer, R. and Haslam, E. (1986). Fast atom bombardment mass spectrometry of polyphenols (*syn.* vegetable tannins). *Biomedical and Environmental Mass Spectrometry,* **13,** 449-468.

92. Decosterd, L.A.;; Dorsaz, A.C. and Hostettman, K. (1987). Application of semi-preparative high performance liquid chromatography to difficult natural product separations. *Journal of Chromatography* **406**, 367-373.

93. Putman, L.J. and Butler, L.G. (1985). Fractionation of condensed tannins by counter-current chromatography. *Journal of Chromatography* **318**, 85-93.

94. Banwart, W.L.; Porter, P.M.; Granato, T.C. and Hassett, J.J. (1985). HPLC separation and wavelength area ratios of more than 50 phenolic acids and flavonoids. *Journal of Chemical Ecology* **11**, 383-395.

95. Hatano, T.; Yoshida, T. and Okuda, T. (1988). Chromatography of tannins III. Multiple peaks in high-performance liquid chromatography of some hydrolyzable tannins. *Journal of Chromatography* **435,** 285-295.

96. Daigle, D.J. and Conkerton, E.J. (1988). Analysis of flavonoids by HPLC: An update. *Journal of Liquid Chromatography* **11**, 309-325.

97. Vande Casteele, K.; Geiger, H.; De Loose, R. and Van Sumere, C.F. (1983). Separation of some anthocyanidins, anthocyanins, proanthocyanidins and related substances by reversed-phase high-performance liquid chromatography. *Journal of Chromatography* **259,** 291-300.

98. Singleton, V.L. and Rossi. J.A. Jr. (1965). Colorimetry of total phenolics with phosphomolybdic-phosphotungstic acid reagents. *American Journal of Enology and Viticulture* **16,** 144-158.

99. Price, M.L.; Van Scoyoc, S. and Butler, L.G. (1978). A critical examination of the vanillin reaction as an assay for tannin in sorghum grain. *Journal of Agricultural and Food Chemistry* **26,** 1214-1218.

100. Walter, W.M. and Purcell, A.E. (1979). Evaluation of several methods for analysis of sweet potato phenolics. *Journal of Agricultural and Food Chemistry* **27,** 942-946.

101. King, H.G.C. and Pruden, G. (1970). Lower limits of molecular weights of compounds excluded from Sephadex G-25 eluted with aqueous acetone mixtures. Application of the results to the separation of the components of tannic acid. *Journal of Chromatography* **52,** 285-290.

102. Reed, J.D.; Horvath, P.J.; Allen, M.S. and Van Soest, P.J. (1985). Gravimetric determination of soluble phenolics including tannins from leaves by precipitation with trivalent ytterbium. *Journal of the Science of Food and Agriculture* **36,** 255-261.

103. Porter, L.J. (1974). Extractives of *Pinus radiata* bark. 2. Procyanidin constituents. *New Zealand Journal of Science* **17,** 213-218.

104. Stafford, H.A. and Cheng, T.Y. (1980). The procyanidins of Douglas Fir seedlings, callus and cell suspension cultures derived from cotyledons. *Phytochemistry* **19,** 131-135.

105. Galletti, C.G. and Self, R. (1986). The polyphenols (*Syn.* vegetable tannins) of grape skins and pressed fruit residues. *Annali di Chimica* **76,** 195-211.

106. Porter, L.J.; Hrstich, L.N. and Chan, B.G. (1986). The conversion of procyanidins and prodelphinidins to cyanidin and delphinidin. *Phytochemistry* **25,** 223-230.

107. Swain, T. and Hillis, W.E. (1959). The phenolic constituents of *Prunus domesticus* 1. The qualitative analysis of phenolic constituents. *Journal of the Science of Food and Agriculture* **10,** 63-68.

108. Burns, R.E. (1963). Methods of tannin analysis for forage crop evaluation. Georgia Agricultural Experiment Stations. Technical Bulletin N.S.32.

109. Folin, O. and Ciocalteau, V. (1927). On tyrosine and tryptophane determination in proteins. *Journal of Biology and Chemistry* **73,** 627-650.

110. Price, M.L. and Butler, L.G. (1977). Rapid visual estimation and spectrophotometric determination of tannin content of sorghum grain. *Journal of Agricultural and Food Chemistry* **25,** 1268-1273.

111. Sarkar, S.K. and Howarth, R.E. (1976). Specificity of the vanillin test for flavanols. *Journal of Agricultural and Food Chemistry* **24,** 317-320.

112. Bate-Smith, E.C. (1972). Detection and determination of ellagitannins. *Phytochemistry* **11,** 1153-1156.

113. ALCA (American Leather Chemists Association) (1956). Sub-committee report. *Journal of the American Leather Chemists Association* **51**, 353.

114. Hoff, J.E. and Singleton, K.I. (1977). A method for determination of tannins in foods by means of immobilised protein. *Journal of Food Science* **42**, 1566-1569.

115. Asquith, T.N. and Butler, L.G. (1985). Use of dye-labelled protein as spectrophotometric assay for protein precipitants such as tannin. *Journal of Chemical Ecology* **11**, 1535-1544.

116. Markham, K.R. (1982). Techniques of Flavonoid Identification. Academic Press, London.

117. Geiger, H. (1985). The identification of phenolic compounds by colour reactions. *Annual Proceedings of the Phytochemical Society of Europe* **25**, 45-56.

118. Harborne, J.B. and Mabry, T.J. (1982). *The Flavonoids: Advances in Research.* Chapman & Hall, London.

119. Mabry, T.J.; Markham, K.R. and Thomas, M.B. (1970). The systematic identification of flavonoids. Springer-Verlag, New York.

120. Foo, L.Y. and Porter, L.J. (1981). The structure of tannins of some edible fruits. *Journal of the Science of Food and Agriculture* **32**, 711-716.

121. Gupta, R.K. and Haslam, E. Plant proanthocyanidins. Part 5. Sorghum polyphenols. *Journal of the Chemical Society, Perkin* **I**, 892-896.

122. Gujer, R.; Magnolato, D. and Self, R. (1986). Glycosylated flavonoids and other phenolic compounds from sorghum. *Phytochemistry* **25**, 1431-1436.

123. Lorenz, K. (1983). Tannins and phytate content in Proso millets (*Panicum miliaceum*). *Cereal Chemistry* **60**, 424-426.

124. Mulkay, P.; Touillaux, R. and Jerumanis, J. (1981). Proanthocyanidins of barley: separation and identification. *Journal of Chromatography* **208**, 419-423.

125. Bate-Smith, E.C. (1973). Tannins of herbaceous Leguminosae. *Phytochemistry* **12**, 1809-1812.

126. Goplen, B.P.; Howarth, R.E.; Sarkar, S.K. and Lesins, D. (1980). A search for condensed tannins in annual and perennial species of *Medicago, Trigonella* and *Onobrychis*. *Crop Science* **20**, 801-804.

127. Sarkar, S.K.; Howarth, R.E. and Goplen, B.P. (1976). Condensed tannins in herbaceous legumes. *Crop Science* **16**, 543-546.

128. Sinha, R.P. and Nath, K. (1982). Effect of urea supplementation of nutritive value of deoiled sal-meal in cattle. *Indian Journal of Animal Science* **52**, 1165-1169.

109

129. Glyphis, J.P. and Puttick, G.M. (1988). Phenolics is some Southern African mediterranean shrubland plants. *Phytochemistry* **27,** 743-751.

130. Reed, J.D. (1986). Relationships among soluble phenolics, insoluble proanthocyanidins and fiber in East African browse species. *Journal of Range Management* **39,** 5-7.

131. Kumar, R. and Horigome, T. (1986). Fractionation, characterisation and protein-precipitating capacity of the condensed tannins from *Robinia pseudo acacia* L. leaves. *Journal of Agricualtural and Food Chemistry* **34,** 487-489.

132. Malan, E. and Pienaar, D.H. (1987). (+)-Catechin-galloyl esters from the bark of *Acacia Gerrardii*. *Phytochemistry* **26,** 2049-2051.

133. Sun, D.; Wong, H. and Foo, L.Y. (1987). Proanthocyanidin dimers and polymers from *Quercus dentata*. *Phytochemistry* **26,** 1825-1829.

134. Schaffert, R.E.; Lechtenberg, V.L.; Oswalt, D.L.; Axtell, J.D.; Pickett, R.C. and Rhykerd, C.L. (1974). Effect of tannin on *in vitro* dry matter and protein disappearance in sorghum grain. *Crop Science* **14,** 640-643.

135. Armstrong, W.D.; Featherston, W.R. and Rogler, J.C. (1973). Influence of methionine and other dietary additions on the performance of chicks fed bird resistant sorghum grain diets. *Poultry Science* **52,** 1592-1599.

136. McGarth, R.M.; Kaluza, W.Z.; Daiber, K.H.; van der Riet, W.B. and Glennie, C.W. (1982). Polyphenols of sorghum grain, their changes during malting, and their inhibitory nature. *Journal of Agricultural and Food Chemistry* **30,** 450-456.

137. Cummins, D.G. (1971). Relationships between tannin content and forage digestibility in sorghum. *Agronomy Journal* **63,** 500-502.

138. Reichert, R.D.; Fleming, S.E. and Schwab, D.J. (1980). Tannin deactivation and nutritional improvement of sorghum by anaerobic storage of H_2O-, HCl-, or NaOH-treated grain. *Journal of Agricultural and Food Chemistry* **28,** 824-829.

139. Sripad, G. and Rao, M.S.N. (1987). Effect of methods to remove polyphenols from sunflower meal on the physicochemical properties of the proteins. *Journal of Agricultural and Food Chemistry* **35,** 962-967.

140 Butler, L.G. (1982). Polyphenols and their effects on sorghum quality. In *Proceedings of the International Symposium on Sorghum Grain Quality*. L.W. Rooney and D.S. Murty (eds.). ICRISAT, India. pp 294-311.

13C-NMR SPECTROSCOPY OF LIGNINS AND LIGNOCELLULOSIC MATERIALS

R. FRÜND AND H.-D. LÜDEMANN

Institut für Biophysik und Physikalische Biochemie, Universität Regensburg, Postfach 397, 8400 Regensburg, FRG

SUMMARY

The conditions for obtaining quantitative high resolution ^{13}C-NMR spectra in solution and in the solid state (CPMAS) are evaluated. The relevant time constants for the relaxation of the spin system are determined and discussed. These time constants are determined at two different fields (2.3 and 7.0 Tesla). The quantitative data are presented. It is shown that the combination of solution and solid state spectra permits an estimate of the total lignin and carbohydrate content of the native lignocellulosic material.

INTRODUCTION

The chemical heterogeneity and the difficulties in completely solubilising lignin and lignocellulosic compounds by non-destructive chemical procedures makes the application of most spectroscopic and physical methods to the quantitative description of lignin and lignocellulosic compounds virtually impossible.

Amongst the available spectroscopic and physical methods only high resolution NMR techniques, which can be applied in solution and in the solid state, readily yield qualitative information about structural details of these heterogeneous biopolymers in their native state. In order to overcome the inherent low sensitivity of ^{13}C-NMR experiments all modern instruments apply pulse techniques and obtain the frequency spectra from a Fourier transformation of the accumulated free induction decays. If this technique is applied in a routine fashion it permits in the complex mixtures of these polymers only the positive and qualitative assignment of the structural elements. This is because the large chemical and structural diversity of the various fractions of these materials can lead to a wide range of spin-lattice and spin-spin relaxation times, which can severely distort the intensities observed in the different ranges of chemical shift. After some early work on the qualitative assignment of ^{13}C signals in solutions of various Björkman lignins,[1-3] their low molecular weight building blocks and synthetic lignin-like polymers[5], our group has concentrated on the quantitative characterisation of lignocellulosic material and has initiated a program to evaluate the conditions for obtaining quantitative information from ^{13}C-NMR spectra. Most of the exploratory work has been performed on humic materials and whole soils. However, initial studies of unmodified wheat-straw and straw-partially digested by fungi show that the NMR-technical problems encountered with these materials are very similar.

RESULTS AND DISCUSSION

The high resolution solid state spectra were obtained with the CPMAS-technique[6,7] at 25.1 MHz and 75.4 MHz in Bruker MSL spectrometers at a spinning rate of 4 KHz. Some typical examples are given in Figure 1. It is customary to separate the regions of chemical shift according to the scheme given in Table 1 and to derive from the areas under these ranges the composition of the samples. In the following it will be shown under which experimental conditions this procedure can lead to quantitatively reliable data. The integration of a variety of ^{13}C spectra from lignocellulosic compounds is compiled in Table 1. In these spectra the relative intensity I of the signals is influenced by the time constants of the spin system and the pulse program. The spin system is mainly described by the cross polarisation time T_{CH} which characterizes the time needed fror the build up of the ^{13}C magnetisation and the proton spin lattice relaxation time in the rotating frame $T_1\rho_H$ which gives the time range, during which the protons can be kept in the spin locked state during a cross polarisation experiment. Provided the repetition time of the experiment is long, compared to all other time constants including, the spin lattice relaxation time of the protons T_{1H}, the observed intensity of a given probe is influenced only by the experimental contact time t, the duration of the Hartmann-Hahn matching between the proton- and the ^{13}C spin system. Under these conditions the intensity I_t observed after a contact time t is a function of T_{CH} and $T_1\rho_H$[6,7].

Equation 1 $\qquad I_t = I_o (T_{CH}, T_1\rho_H, t)$

I_o being directly proportional to the number of nuclei in the probe. In the case $T_{CH} \ll T_1\rho_H$ equation 1 simplifies for $t \gg T_{CH}$ to:

Equation 2 $\qquad I_t = I_o \exp(-t/T_1\rho_H)$

$T_1\rho_H$ can thus be determined directly from a plot of log I_t versus t. Figures 2 and 3 show some typical results. Figure 2 gives the intensities observed for the most intense peaks of the four areas indicated in Table 1 for the complete humic extract from a German eutric gleysol. This compound contains a relatively high concentration of unpaired electrons and it was thus to be expected that these would influence all relaxation times observed. The initial increase of the intensity at small contact times is a measure of T_{CH}. It is obvious that there are significant differences found for the four compounds (see insert in Figure 2). However, for all carbons, the condition $T_{CH} \ll T_1\rho_H$ is fulfilled. The same holds for the results obtained from the wheat straw sample shown in Figure 3.

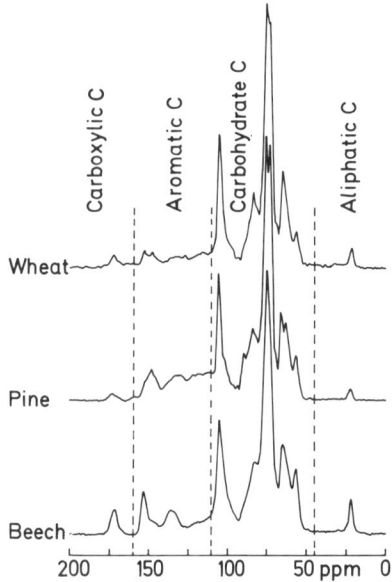

Figure 1. ¹³C-CPMAS spectra of milled beech wood, pine wood and wheat straw. The vertical bars at 46, 110, 160 ppm give the separation of the spectral region into aliphatic-, carbohydrate; aromatic- and carboxylic-carbons.

TABLE 1. Composition of a variety of unmodified milled lignocellulosic materials as obtained from CPMAS spectra at 25.1 MHz. (Spin rate: 4 kz; contact time: 1 ms; recycle time: 5 ms)

Plant	Concentration (%)			
	Carboxylic-C 210-160 ppm	Aromatic-C 160-110 ppm	Carbohydrate-C 110-4 ppm	Aliphatic-C 46-0 ppm
Wheat straw	2.1	10.7	83.8	3.4
Barley straw	1.5	11.6	83.9	3.0
Oat straw	2.5	13.4	80.7	3.3
Rye straw	3.3	12.6	80.0	4.0
Hay sample	5.6	12.4	72.5	9.5
Beech	2.0	13.2	82.5	2.3
Mahogany	1.8	24.2	70.6	3.2
Pine	0.7	19.0	78.9	1.4
Spruce	0.9	18.2	79.0	1.9

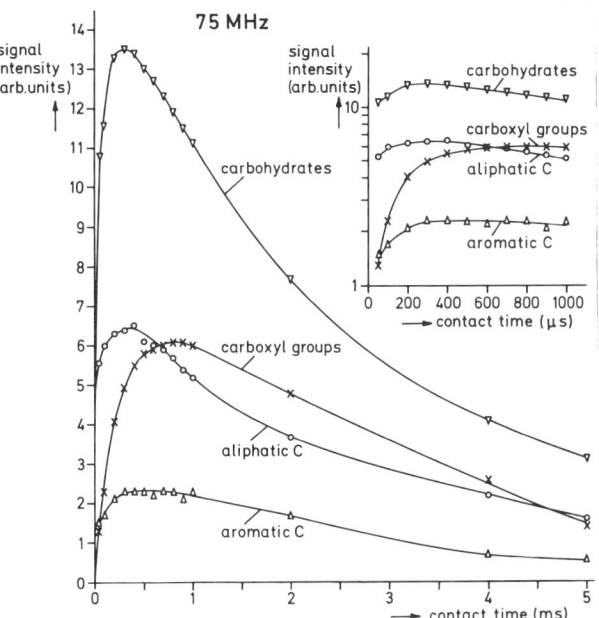

Figure 2. Signal intensity of the ^{13}C CPMAS spectrum of a sodium hydroxide extract of humic material from an euteric gleysol as function of contact time. Insert: Logarithmic plot of the intensities for very short contact times.

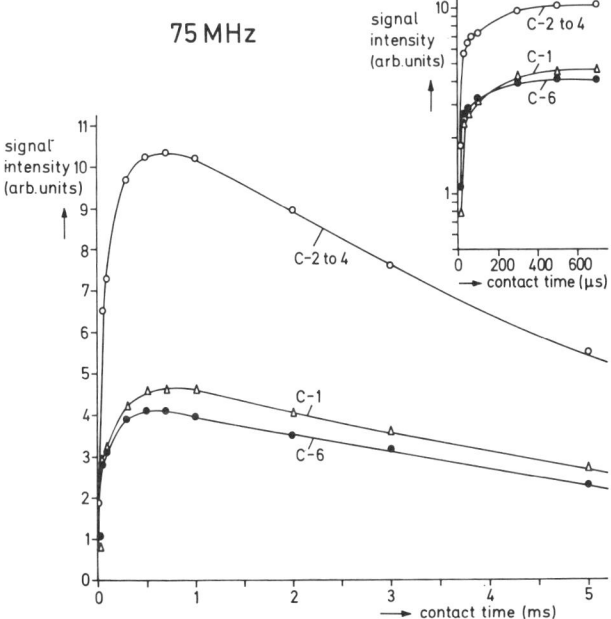

Figure 3. Signal-intensity of the ^{13}C CPMAS spectrum of wheat straw as function of contact time. Insert: Logarithmic plot of the intensities for very short contact times.

The decrease of I_t at $t > 1$ ms is linear in a log I_t versus t plot. From the slope of this line $T_{1\rho H}$ can be determined. The results of this analysis are given in Table 2. Attempts to fit the complete individual curves of the Figures 2, 3 with the $T_{1\rho H}$ data of Table 2 and one T_{CH} relaxation time by application of equation 1 in its extended form[6,7] failed. The initial increases in the intensity are obviously described by a distribution of T_{CH} values. The data presented would permit a recycle time for the cross polarisation experiment of at most 100 ms. However, the complete spin polarisation can only be obtained if the experiment is started from a proton spin system in thermal equilibrium. The approach to thermal equilibrium is described by the spin lattice relaxation time T_{1H} of the protons. It is possible and advantageous to measure T_{1H} in the ^{13}C-spectra with a special pulse sequence, similar to an inversion recovery experiment in the CPMAS-mode[6,7]. These experiments were performed for the humic extract and the unmodified wheat straw samples (Table 3). From these data it becomes obvious, that the presence of the paramagnetic centres is most dramatically seen in T_{1H}. For practical spectroscopy the data imply that, for humic material, the recycle time may be as low as 100 ms while for unmodified lignocellulosic material this time must be at least 4000 ms. Consequently the total time needed for obtaining a good spectrum is much longer for the native compounds. It is to be suspected, that the presence of paramagnetic centres in a complex organic material will lead for a fraction of the carbons to very short spin-spin-relaxation times T_2[5,6]. These carbons could become unobservable in the spectra and invalidate attempts to extract quantitative information from the CPMAS spectra. In order to obtain a reliable estimate of this effect, the average carbon content obtained from the solution spectra of 10 humic materials at very long aquisition delays were compared to CPMAS spectra obtained at contact and recycle times permitting a quantitative evaluation. These data are compiled in Table 4. The agreement between these 3 sets of data is better than expected. The relative intensities found for these three approaches agree to better than ± 2%. With the limited data set available, it is premature to speculate about possible physical reasons for the differences observed.

The concentrations derived in Table 1 for the various classes of carbon appear thus to be representative for the composition of the native lignocellulosic material. The limited resolution of the solid state spectra however does not allow separation in the chemical range between 100 and 110 ppm aromatic carbons or the C-2 and C-6 position of syringyl units which should be very prominent in hardwood lignins from the anomeric C-1 of carbohydrates. In solution and solid state spectra of pure lignins (1,6) the syringyl units do show a very pronounced signal at ~ 105 ppm and also the corresponding C-3/C-5 signals at ~ 155 ppm. Both these signals must be missing in lignins from conifers lacking syringyl units. Separate integration of the area between 160 and 150 ppm gives, for beech and mahogany, approximately 4% of the total intensity while only 1.5 to 2% are seen in the spruce-, pine- and straw lignins. It is thus to be expected that the carbohydrate concentration is overestimated and correspondingly the aromatic carbon content underestimated by these amounts.

TABLE 2. $T_{1\rho_H}$ at 300 K for humic material from a eutric gleysol and for the cellulose carbons of wheat straw obtained from the intensity of the ^{13}C-CPMAS spectra as function of contact time and observed frequency. For the wheat only the intense signals in the region of the carbohydrate carbons could be studied. Their assignment is indicated.

Frequency	Humic material			
	Carboxyl-C	Aromatic-C	Carbohydrate-C	Aliphatic-C
75.4 MHz	3.4 ms	3.0 ms	3.1 ms	3.3 ms
25.1 MHz	3.2 ms	3.9 ms	2.7 ms	3.7 ms
Frequency	Wheat straw			
	C-1 (105 ppm)	C-6 (65 ppm)	C-2 to 4 (75 ppm)	
75.4 MHz	7.6 ms	7.5 ms	6.9 ms	

TABLE 3. Spin lattice relaxation times T_{1H} of the protons in humic materials and wheat straw.

Material	Frequency (MHz)	T_{1H} (ms)
Humic extract	300	10-20
Humic extract	100	3- 6
Wheat straw	300	800-1300

Lignins however carry at least 4 aliphatic carbons per benzene ring which yield signals in the chemical shift range of the carbohydrate carbons in the form of their propane sidechain and as methoxy groups. The total lignin carbon content is thus at least a factor of 1.7 higher than the corrected aromatic carbon content, and the carbohydrate content must be reduced by the corresponding amount. This crude estimate suggests a lignin content of approximately 21% and carbohydrate content of approximately 74% for wheat straw. For spruce, beech and mahogany the corresponding values for lignin are 33%, 29% and 47% respectively while the carbohydrate values are reduced to 65%, 66% and 57%. These estimates are within the range of the commonly accepted value and this procedure shows, that more detailed information can be obtained from a combination of the generally well resolved solution spectra with quantitative CPMAS spectra.

TABLE 4. Composition of humic materials from soil determined from solution spectra at 75.4 MHz and from CPMAS spectra at 25.1 and 75.4 MHz

	Concentration (%)			
	Carboxyl-C	Aromatic-C	Carbohydrate-C	Aliphatic-C
Solution Spectra 75MHz	12.0	17.7	49.0	21.3
CPMAS 25 MHz	12.2	17.2	51.9	18.7
CPMAS 75 MHz	12.5	13.6	51.9	22.0

Finally we want to present our first application of the CPMAS technique to a wheat straw treated with the fungus *Bjerkandera adusta*. The composition of the starting materials is shown in Table 1. After 42 days fermentation the carboxyl carbon content was increased to 3.0%, the aromatic carbon content reduced to 9.0% and the aliphatic residues increased to 4.4%. Although it is premature to discuss these effects quantitatively they are outside the limits of experimental error and corroborated by spectra taken at intermediate fermentation times. Other obvious effects of fermentation are the specific losses of approximately 30% of the intensity of the peak at 61 ppm, which is assigned to methoxyl groups of aromatic lignin units, corroborating the reduction of the content of aromatic carbons.

Acknowledgement

A part of this work was supported by a grant from the German Bundesministerium für Forschung und Technologie.

REFERENCES

1. Lüdemann, H.-D. and Nimz, H. (1973). Carbon-13 Nuclear Magnetic Resonance Spectra of Lignins. *Biochemical and Biophysical Research Communications* **52**: 1162.

2. Lüdemann, H.-D. and Nimz, H. (1974). Kohlenstoff-13-Kernresonanzspektren von Ligninen, 1. Chemische Verschiebungen von monomeren und dimeren Modellsubstanzen. *Makromolekulare Chemie* **175**, 2393.

3. Lüdemann, H.-D. and Nimz, H. (1974). Kohlenstoff-13-Kernresonanzspektren von Ligninen, 2. Buchen- und Fichten-Björkman-Lignin. *Makromolekulare Chemie* **175**, 2409.

4. Nimz, H. and Lüdemann, H.-D. (1974). Kohlenstoff-13-Kernresonanzspektren von Ligninen, 5. Oligomere Ligninmodellsubstanzen. *Makromolekulare Chemie* **175**, 2577.

5. Nimz, H.; Morharab, I. and Lüdemann, H.-D. (1974). Kohlenstoff-13-Kernresonanzspektren von Lignin, 3. Vergleich von Fichtenlignin mit künstlichem Lignin nach Freudenberg. *Makromolekulare Chemie* **175**, 2563.

6. Wilson, M.A. (1987). *NMR Techniques and Applications in Geochemistry and Soil Chemistry*, Pergamon Press, Oxford/New York.

7. Mehring, M. (1983). *Principles of High Resolution NMR in Solids*. Springer Verlag, Berlin.

LIGNIN CHARACTERIZATION OF WHEAT STRAW SAMPLES AS DETERMINED BY CHEMICAL DEGRADATION PROCEDURES

C. LAPIERRE, D. JOUIN and B. MONTIES

Laboratoire de Chimie Biologique, INRA, INA-Paris-Grignon, F78850 Thiverval Grignon (France)

SUMMARY

Extractive-free wheat straw, and the corresponding samples issued from its alkaline treatment - i.e. saponified wheat straw and alkali-soluble wheat lignin- were submitted to three chemical procedures of lignin characterization: nitrobenzene oxidation, acidolysis and thioacidolysis. The various performances of the three techniques were comparatively evaluated. The thioacidolysis procedure was found to have higher performances with regards to the two following criteria: the prominent alkyl aryl ether lignin structures were the almost exclusive source for the recovered degradation products; these products were obtained with a high yield. These advantages make the thioacidolysis procedure an unambiguous way to characterize grass lignins, even when present in low amounts.

INTRODUCTION

Many studies have been devoted to the characterization of lignin and phenolic acids in the cell wall of forages, as these compounds are known to affect detrimentally their digestibility.[4,16,19,21] However detailed structural data about grass lignins, as opposed to wood lignins, is still very scarce.

The structure and composition of lignins is currently determined by the analysis of their degradation products, such as those recovered from the classical nitrobenzene oxidation procedure applied to feedstuffs.[16] The targets of the main available degradation techniques are essentially the uncondensed alkyl aryl ether bonds established between the p-hydroxyphenyl (H), guaiacyl (G) and/or syringyl (S) lignin building units (Figure 1a). Structures such as those illustrated in Figure 1a are the most characteristic ones met in grass lignins.

Recently, a new lignin depolymerization procedure, thioacidolysis (i.e. solvolysis in dioxane-ethanethiol, 9/1, v/v, 0.2 M boron trifluoride etherate) was investigated.[8-11] It was shown to specifically cleave the alkyl aryl ether bonds in lignins and to produce, with a high reaction yield, the main thioethyl products shown in Figure 1b.

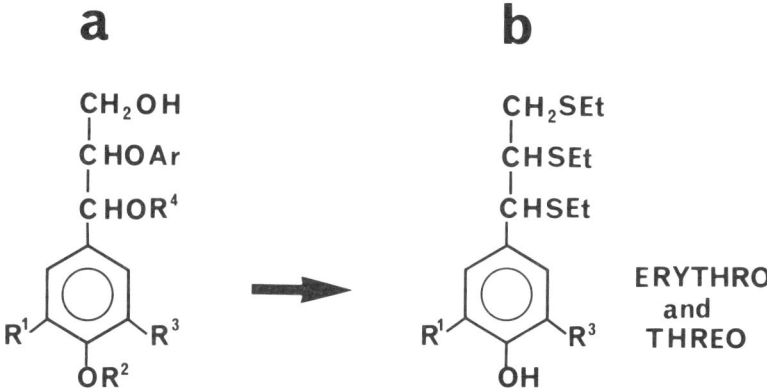

Figure 1. Thioacidolysis of uncondensed alkyl aryl ether structures. $R^1 = R^3 = H$ in p-hydroxyphenyl units; $R^1 = OMe$ and $R^3 = H$ in guaiacyl ones; $R^1 = R^3 = OMe$ in syringyl ones; $R^2 = H$ or aliphatic C in lignin; $R^4 = H$ or Ar.

The main objective of the present study was to test the applicability and performances of the thioacidolysis procedure applied to wheat samples, as compared to the nitrobenzene oxidation and the acidolysis techniques. The two latter methods are those most frequently carried out by lignin chemists. In addition, it was found interesting to examine the fate of p-coumaric acid (PC) and ferulic acid (FE) units, in the course of the degradation, because their presence is a characteristic feature of grass cell walls.[4,14,19]

MATERIALS AND METHODS

Samples

Wheat straw (*Triticum aestivum*, c.v Capitole) was used to prepare an extractive-free wheat cell wall residue, (CW) by successive extractions of the milled straw with ethanol:toluene (1:2. v/v), ethanol and then water in a Soxhlet apparatus.

Wheat saponified residue, (SR) and alkali-soluble lignin fraction, (AL), were prepared from CW according to the method described by Scalbert *et al.*[18] The Klason lignin content of CW and SR was 19.5 and 13.8%, respectively. The yield of AL, expressed as weight percentage of Klason lignin in CW, was 24.5%.

Degradation procedures

Nitrobenzene oxidation: 2M aqueous NaOH solution (5 ml) and nitrobenzene (0.5 ml) were added to samples of CW or SR (25 mg) or to AL (10 mg) in a Teflon vial enclosed within a stainless steel autoclave. The autoclaves were heated in an oil bath 3 hours at 160°C, with gentle shaking. The cooled reaction mixture was diluted with ~ 10 ml H_2O and extracted with 3x50 ml CH_2Cl_2, to eliminate degradation products from nitrobenzene. The aqueous residue was then adjusted to pH 1-2 with concentrated HCl and extracted with 3x50 ml CH_2Cl_2. The HPLC internal standard (anisaldehyde 2 mg) was then added to the organic extracts, which were evaporated to dryness at reduced pressure before being redissolved in 1 ml CH_2Cl_2. The *p-hydroxybenzaldehyde, vanillin and syringaldehyde,* respectively recovered from H, G and S units with various sidechains, were estimated by reverse phase HPLC, using a procedure similar to that previously described by Scalbert *et al.*[18]

Acidolysis: They were performed from 15 mg CW or SR or from 5 mg AL, according to a method previously reported.[7] The guaiacyl and syringyl main recovered products were quantified by gas chromatography of their trimethylsilylated (TMS) derivatives, with tetracosane as internal standard and on a 5% phenyl-, 95% methyl-silicone-fused silica capillary column (0.3 µm film thickness, 25 m x 0.2 mm i.d., AML Chromato.). Acidolysis products were, in all cases, identified by mass spectroscopy.

Thioacidolysis: All reagents were of pro analysis grade. CW or SR (15 mg) or AL (5 mg) were mixed with 10 ml of dioxane:ethanethiol mixture (9:1, v/v) and 0.2 M BF_3 etherate in a tube fitted with a teflon-lined screwcap. The thioacidolysis was then allowed to proceed under nitrogen for 4 hours at 100°C (oil bath) and with occasional shaking. The cooled reaction mixture was diluted with ~ 10 ml H_2O, adjusted to pH 3-4 with aqueous 0.4 M $NaHCO_3$ and extracted with 3 x 50 ml CH_2Cl_2. The organic extracts, with tetracosane and hexacosane (0.5 mg each) added as GC internal standards, were dried over Na_2SO_4 and concentrated by film evaporation. The final residue was dissolved in ~ 1 m CH_2Cl_2. Silylation, gas chromatography (GC) and gas chromatography-mass spectrometry (GC-MS) analyses were as previously described.[9,10] Ferulic acid was submitted to a similar thioacidolysis procedure in order to identify its degradation products.

RESULTS AND DISCUSSION

In the present paper, emphasis is placed on the comparison of the degradation methods and not on the structural modification caused by the alkali treatment of wheat lignin.

Nitrobenzene oxidation data are shown in Table 1. For the purpose of comparison between methods, yields are expressed in micromoles of degradation products per gram of lignin. The H, G and S compounds, which are quantified in Table 1, are *p*-hydroxybenzaldehyde, vanillin and syringaldehyde, respectively. GC-MS analysis, showed that other degradation products could be ignored because of their relative unimportance.

TABLE 1. Yields of the alkaline nitrobenzene oxidation products recovered from wheat samples, for the H, G and S series. Yields are expressed as μmoles per g of Klason lignin for CW or SR and per g of alkali soluble lignin for AL.

Fraction	H	G	S	H/G	S/G
Cell wall residue (CW)	542	1036	940	0.52	0.92
Saponified residue (SR)	340	1192	1220	0.29	1.02
Alkali soluble lignin (AL)*	220	858	913	0.26	1.06

* Yields are not corrected for carbohydrate contaminants in AL fraction. Standard errors between duplicate experiments were less than 10%.

According to the data in Table 1, original wheat lignin appeared as a typical H-G-S lignin, as the relative distribution (% on a molar basis) of the H, G and S degradation products from CW was 22:41:37 respectively. However, the relative importance of H compounds drastically decreased in SR and AL. Such behaviour has been extensively studied by Higuchi and his coworkers (1967). These authors found that more than half of *p*-hydroxybenzaldehyde recovered from various grass samples was due to *p*-coumaric acid and not to lignin. Although present in smaller amounts than *p*-coumaric acids, the presence of ferulic acid suggest that the origin of vanillin may be similarly ambiguous, as ferulic acid is degraded to vanillin with a good reaction yield.[22] Therefore, the changes in the S/G or H/G ratio caused by the alkaline treatment and observed in Table 1 are not only related to structural modifications of lignins, but also to losses of PC and FE units within the wheat samples. Accordingly, it appears that the nitrobenzene oxidation procedure, which has the advantage of being a routine procedure, is not capable of producing a direct and unambiguous characterization of grass lignins, because of the presence of PC and FE units in the grass cell walls.

In contrast to nitrobenzene oxidation, acidolysis and thioacidolysis preserve the characteristic phenylpropane skeleton of the degraded lignin structures in the recovered products. These degraded structures are those depicted in Figure 1a.

The complexity of the acidolysis mixture from wheat samples is shown in the chromatogram of Figure 2. Identification of recovered products is shown in Table 2.[2,6,7] This complexity is one of the drawbacks of the acidolysis procedure, which makes it less useful for routine analysis. Some minor peaks of Figure 2 have been assigned to chlorinated syringyl structures, from their mass fragmentation patterns and from model compound experiments.[6] Analogous H and G compounds have never been found. These S chlorinated derivatives are not taken into account for calculations, as they are of only minor importance. However, they clearly reveal that uncontrolled side reactions occur particularly within S units. Peaks assigned to H compounds (peaks 3, 5, 8, 9 and 17) have a weak relative importance, compared to their parent G and S products. Although these peaks were not determined quantitatively as some of them were not suitably resolved, this result may indicate that the relative importance of H building units in the uncondensed lignin moiety is low. PC and FE acids could be identified in the wheat acidolysate (peaks 15 and 24). Although acidolysis does not liberate quantitatively PC and FE ester- and/or ether-linked units, the contents of recovered PC and FE acids are reported in Table 3, for comparative purposes. It was observed that AL fraction was relatively enriched with FE units as compared to original lignin in CW or residual lignin in SR samples. This result confirms that FE units are more closely associated with the alkali-soluble wheat lignin fraction, in agreement with a more extensive study about wheat lignins.[17] Furthermore, it emphasizes that the S/G ratio shown in Table 1 for AL samples is not a close reflection of lignin composition.

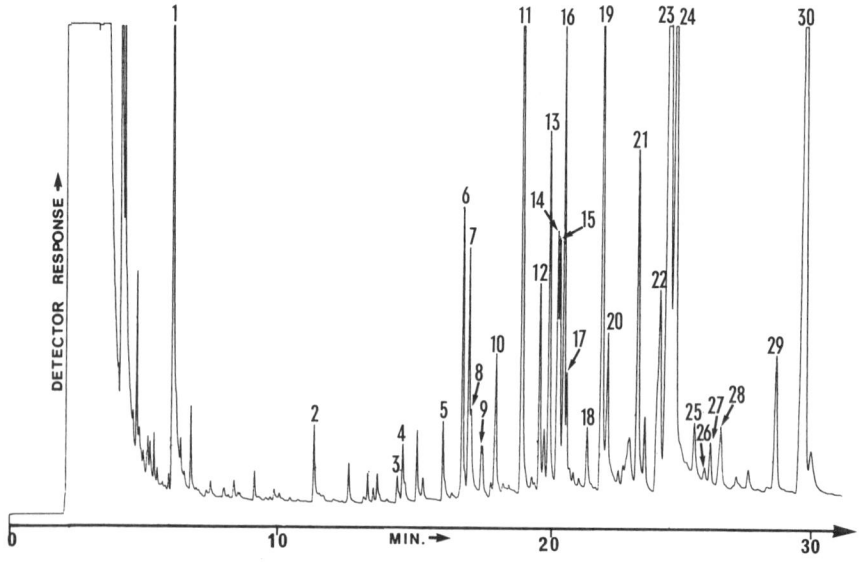

Figure 2. GC chromatogram of TMS acidolysis monomeric products recovered from wheat alkali-soluble lignin fraction. For peak identification, see Table 2.

Data reported in Table 3 show that the acidolysis yields for the G and S compounds are relatively low, compared to the values in Table 1. Such a difference cannot be related only to the fact that the oxidative procedure degrades some further structures than those depicted in Figure 1a. In fact, this difference reveals one of the most severe drawbacks of the acidolysis procedure; notably that during acidolysis, side reactions occur which severely counteract the recovery of lignin degradation products.[13] From studies of G and S appropriate

TABLE 2. Structure of the TMS derivatives from the acidolysis of wheat samples, separated in Figure 2.

Peak	TMS derivative of	TMS form after isomerisation	Molecular ion and prominent fragments m/e (rel. intensity)
1	5-hydroxymethylfurfural	-	M=198(10), 183(100), 169(23), 109(64), 73(100)
2	G-CHO	-	M=224(20), 209(40), 194(100), 73(25)
3	H-CHOH-CO-CH₃	ketonic	M=310(0), 295(10), 267(100), 193(10), 73(90)
4	S-CHO	-	M=254(30), 239(50), 224(100), 73(20)
5	H-CO-CHOH-CH₃	ketonic	M=310(10), 295(5), 193(70), 117(80), 73(100)
6&10	G-CH₂-CHO	Z and E enolic	M=310(90), 295(10), 280(100), 179(10), 73(100)
7	G-CHOH-CO-CH₃	ketonic	M=340(2), 325(12), 297(100), 73(70)
8	H-CO-CO-CH₃	enolic	M=308(20), 293(50), 193(60), 73(100)
9	H-CH₂-CO-CH₂OH	ketonic	M=310, 295, 179, 14;9, 103, 73
	and		
	S-CH=CH=CHO	aldehyde	M=280, 265, 250, 222, 73
11	G-CO-CHOH-CH₃	ketonic	M=340(10), 325(20), 223(100), 117(90), 73(95)
12	S-CHOH-CO-CH₃	ketonic	M=370(2), 355(20), 327(100), 73(90)
13	G-CO-CO-CH₃	enolic	M=338(10), 323(30), 307(50), 292(15), 279(10), 223(30), 73(100)
14&18	S-CH₂-CHO	Z and E enolic	M=340(50), 325(5), 310(60), 73(100)
15	PC	-	M=308(40), 293(50), 249(25), 219(50), 179(10), 139(10), 73(100)
16	G-CH₂-CO-CH₂OH	ketonic	M=340(15), 3255), 209(90), 179(15), 130(10), 73(100)
17	H-CH₂-CO-CH₂OH	enolic	M=382(60), 367(5), 293(10), 192(20), 147(60), 73(100)
19	S-CO-CHOH-CH₃	ketonic	M=370(15), 355(20), 253(100), 223(10), 117(60), 73(100)
20&23	G-CH₂-CO-CH₂OH	Z and E enolic	M=412(50), 297(5), 323(5), 222(5), 192(10), 147(25), 73(100)
21	S-CO-CO-CH₃	enolic	M=368(20), 353(20), 337(75), 253(70), 223(10), 73(100)
22	S-CH₂-CO-CH₂OH	ketonic	M=370(40), 355(10), 239(100), 209(20), 179(5), 103(10), 73(100)
24	FE	-	M=338(90), 323(50), 308(50), 293(50), 279(10), 249(30), 219(20), 73(100)
25&29	S-CHCl-CO-CH₃	Z and E enolic	M=388(40), 390(20), 375(5), 373(10), 360(5), 358(10), 353(20), 323(20), 73(100)
27&28	S-CHOH-CHCl-CH₂OH	erythro and threo	M=478(10), 463(5), 327(100), 239(10), 103(10), 73(100)
30	S-CH₂-CO-CH₂OH	enolic	M=442(50), 427(5), 222(10), 147(20), 73(100)

Structures of peaks (25,29) and (27,28) were tentatively assigned from mass fragmentation patterns model compounds, it was found that S structures had a higher tendency to participate to side reactions than the parent G compounds. It is therefore obvious that the S/G ratio given in Table 3 do not reflect accurately the actual S/G ratio of the lignin degraded structures.

TABLE 3. Yields of the acidolysis products recovered from wheat samples. Yields are expressed as in Table 1. H compounds are not determined owing to unsuitable separation

Fraction	p-Coumaric acid	Ferulic acid	G	S	S/G
Cell wall residue (CW)	87	74	316	206	0.65
Saponified residue (SR)	17	20	417	240	0.58
Alkali-soluble lignin (AL)*	27	122	332	250	0.75

* Yield not corrected for carbohydrate components. Standard errors between duplicate experiments were found lower than 10%

We have subjected the wheat samples to thioacidolysis with the aim of degrading the characteristic lignin structures into specific reaction products and without involving simultaneous side reactions. Figure 3, shows the GC chromatograms of the TMS thioacidolysates from wheat CW and AL. The structure of the main recovered phenylpropane adducts is depicted in Figure 4, except for peaks 1 and 3 which are TMS PC and FE acids, respectively.

The thioacidolysis procedure degrades the lignin structures shown in Figure 1a producing very few monomeric compounds compared to acidolysis. The H, G and S recovered compounds are essentially located in pairs of peaks designated 5, 6 and 8, respectively. The weak shoulders, 7 and 9, have the same origin, via a minor reaction pathway.[8]

Thioacidolysis of PC and FE units present in wheat samples leads to PC and FE acids (peaks 1 and 3) together with their parent Michael addition products (peaks 2 and 4). In agreement with acidolysis data, Figure 3 reveals that AL seems enriched with FE units, as compared to CW.

Thioacidolysis yields of the main H, G and S products (pairs of peaks 5, 6 and 8 in Figure 3) are given in Table 4. For the sake of comparison, results from some wood samples are included. Yields from thioacidolysis are about twice more than that produced by acidolysis. Therefore, they reflect more closely not only the relative content, but also the monomeric composition of the uncondensed alkyl aryl ether structures. Compared to the nitrobenzene oxidation yields for S compounds, yields from thioacidolysis are about 30% lower. This difference may be assigned to the structural specificity of the thioacidolysis procedure. While the structures depicted in Figure 1a are the exclusive source of the thioacidolysis products, the nitrobenzene oxidation technique is capable of degrading additional structures, such as the diarylpropane units.[1]

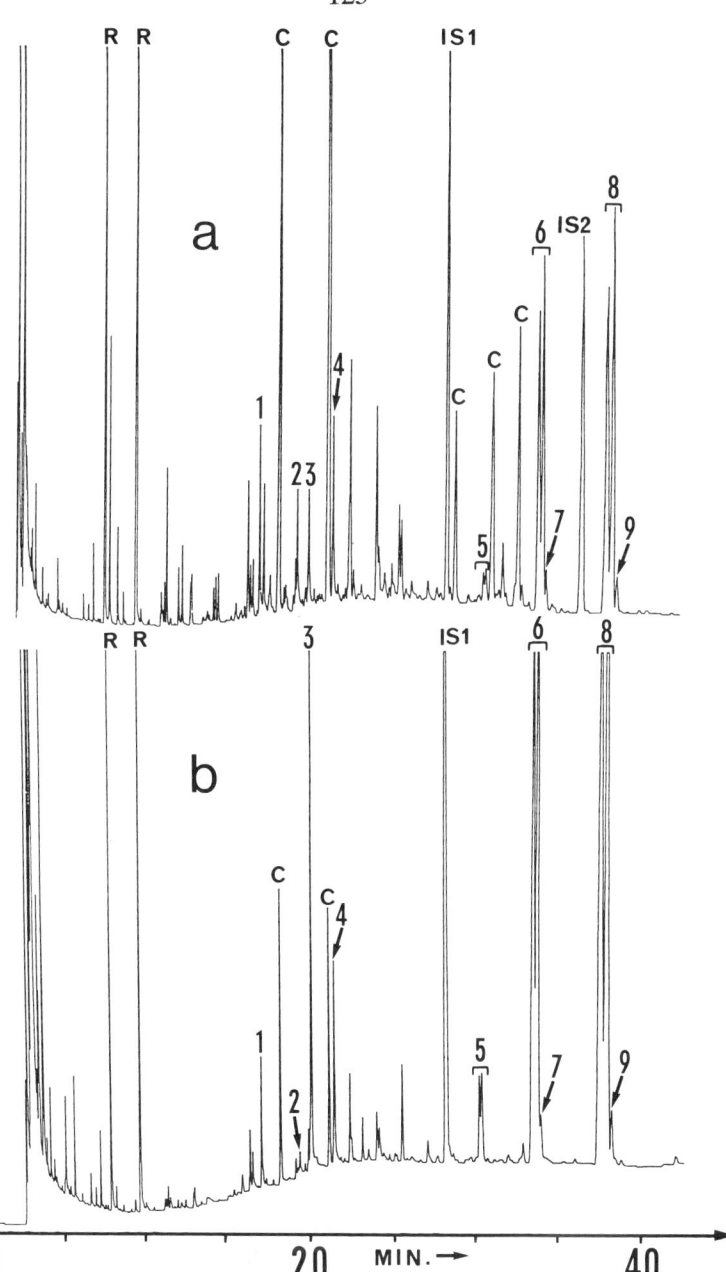

Figure 3. GC chromatograms of TMS thioacidolysis products recovered from a) wheat cell wall residue and b) wheat alkali-soluble lignin fractions. R = peaks from reagent; C = peaks from carbohydrates; 1 and 3: TMS PC and FE acids; 2,4,5-9: see Figure 4; IS1: tetracosane and IS2 : hexacosane (GC internal standards).

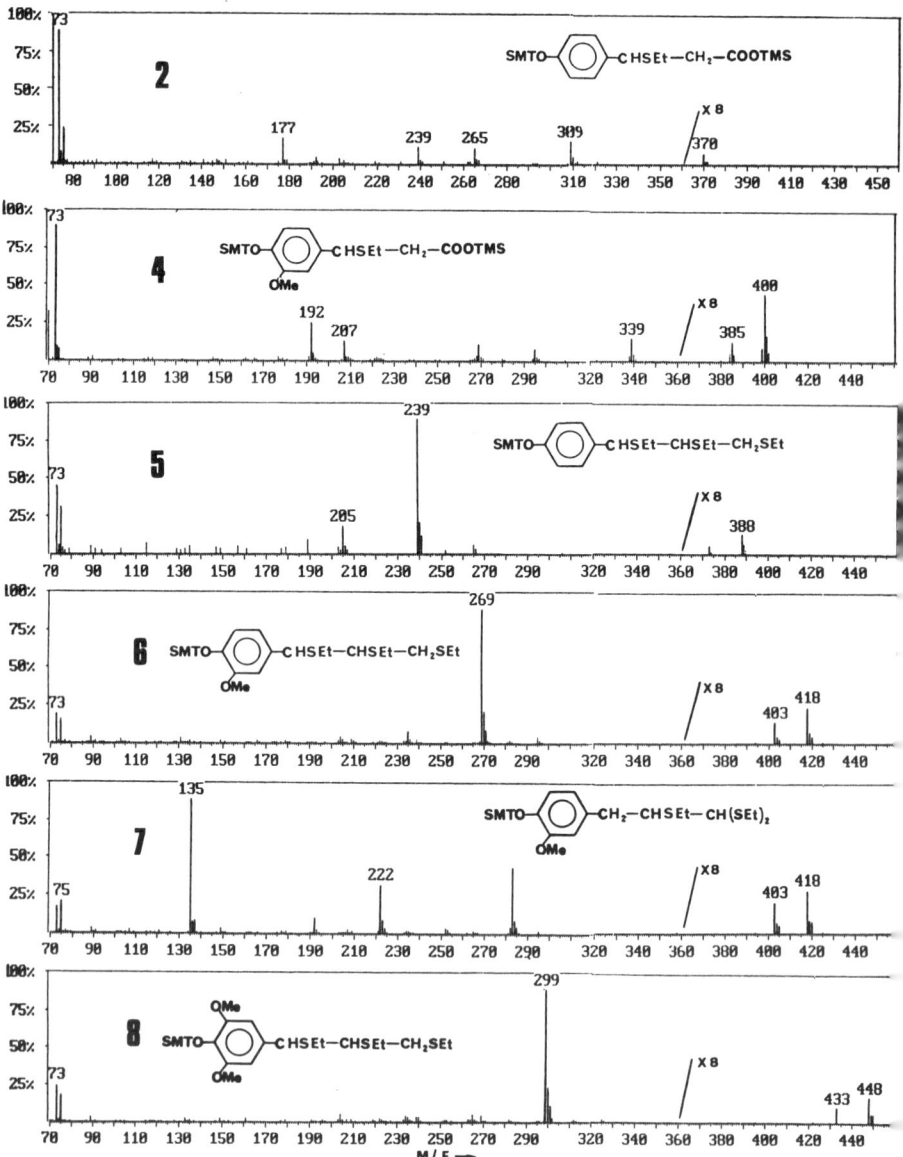

Figure 4. MS identification of peaks observed in Figure 3. Peaks 1 and 3 correspond TMS PC and FE acids; peak 9 is the syringyl parent compound of the guaiacyl peak Erythro and threo isomers of the pairs of peaks 5 or 6 or 8 have similar mass spectra.

The high reaction yield of thioacidolysis, together with the certainty that the quantified H, G and S monomeric compounds originate from lignin, allows the detailed interpretation of results shown in Table 4.

TABLE 4. Yields of thioacidolysis main products recovered from various lignin samples. Yields are expressed as in Table 1.

	H	G	S	Total	H/G	S/G
Poplar wood (CW)	n.d.	763	1187	1950	-	1.56
Pine wood compression wood (CW)	207	934	n.d.	1141	0.22	-
Wheat straw (CW)	53	535	650	1238	0.10	1.21
Wheat straw (SR)	56	684	786	1526	0.08	1.15
Wheat straw (AL)*	53	427	597	1077	0.12	1.40

* Yields not corrected for carbohydrate components. n.d.: not determined owing to negligible relative amounts. Standard deviations between triplicate are lower than 5%

The structure of lignin in wheat CW is closer to that of compression wood lignin than that from hardwood. The higher degree of condensation of grass lignin, compared to hardwood lignin, has been confirmed by a more systematic study of grass (unpublished results) and poplar samples.[6] This result is in agreement with the data reported by Nimz and coworkers, by means of ^{13}C-NMR spectroscopy of isolated wheat lignins.[15]

The S/G ratio of original wheat lignin, in the CW sample, is lower than in the poplar lignin sample. The lower content of syringyl units may be related to the lower amount of alkyl aryl ether bonding, as it was found, in the case of wood, that syringyl units had a higher tendency to be involved in alkyl aryl ether bonds than guaiacyl ones.[3,6] The change in the S/G ratio observed for AL, should be considered with caution, as lignin fragments are certainly lost during the step of AL precipitation. It was found more reasonable to compare CW and SR samples only. From a more systematic study of grass samples, it could be seen that changes in S/G ratio between CW and SR samples were generally weak and unpredictable (Table 4).

The relative importance of recovered H compounds was very low for the three wheat samples (Table 4), as well as for many other grass samples (unpublished results). While the relative importance of H compounds (% on a molar basis) was 4.2, 3.7 and 4.9% of the total yields from wheat CW, SR and AL respectively, the recovered H compounds in pine compression wood lignin accounted for 18% of the total yield. Pine compression wood is known to have a characteristic lignin which, compared to "normal" pine lignin, is an H-G lignin relatively abundant in H building units.[20] This comparison emphasizes that H units have an homogenous and very minor importance in uncondensed wheat lignin structures from CW, SR and AL. One can therefore suppose that H units are scarce in wheat lignins or that they are specifically involved in condensed structures.

CONCLUSION

The results presented here allow us to draw the following conclusions.

In depth investigations of the structure of original or treated lignins are the key step in developing a better understanding of what occurs inside the lignified cell walls of forages when submitted to chemical or biological processes. It is therefore of the utmost importance to look for analytical methods capable of generating detailed structural information about lignins, without any ambiguity

PC and FE units are the major impediment to the accurate characterization of wheat lignin by the nitrobenzene oxidation procedure, as they are degraded into the same compounds as the H and G building units of lignin. In addition, some other phenolic components of the cell walls, such as the tyrosine residue may also interfere.

Contrary to oxidative procedures which causes cleavage in the side chain of phenylpropane units, acidolysis and thioacidolysis procedures cause cleavage of alkyl aryl ether bonds together with the preservation of the characteristic phenylpropane skeleton in the recovered products. This is a major advantage of these solvolytic processes. Although it was fruitfully used some years ago[2], the acidolysis procedure is less effective than thioacidolysis. Degradation products from acidolysis are recovered with a lower yield, particularly compounds which are more susceptible to side reactions resulting in a more complex mixture of end products. In contrast, during thioacidolysis, the characteristic H, G or S lignin structures fragment essentially to only two H, G or S thioethyl phenylpropane diastereoisomers, with a high reaction yield. This yield closely reflects the amount of degraded lignin structures and the relative proportions of the H, G or S diastereoisomers accurately indicate the monomeric composition of these structures. Thioacidolysis results have shown that wheat lignins are more condensed than hardwood lignins and that the relative importance of the H units in their uncondensed moiety is very scarce.

The analysis of the monomeric products recovered from thioacidolysis of grass samples has proved to be a routine procedure. It proved very sensitive as some experiments could be performed with weakly lignified cell wall (quantity less than 15 nanograms of lignin per assay) (unpublished results). Furthermore, detailed analyses of minor lignin degradation products and thioacidolyses of various samples thoroughly treated with diazomethane[12] have provided more structural and sometimes original information about these fascinating macromolecules.

REFERENCES

1. Chang, H.M. and Allan, G.G. (1971). Oxidation. In: *Lignins: Occurrence, Formation, Structure and Reactions*, K.V. Sarkanen and C.H. Ludwig (eds) Wiley-Interscience, New York, pp 433-485.

2. Gellerstedt, G.; Lindfors, E.L.; Lapierre, C. and Monties, B. (1984). Structural changes in lignins during kraft cooking. Part 2-Characterization by acidolysis. *Svensk Papperstidning* **9**: R61-R67.

3. Glasser, W.G.; Barnett, C.A. and Sano, Y. (1983). Classification of lignins with different genetic and industrial origins. *Applied Polymer Symposium* **37**: 441-460.

4. Hartley, R.D. (1972). *p*-Coumaric and ferulic acid components of cell walls of ryegrass and their relationships with lignin and digestibility. *Journal of the Science of Food and Agriculture* **23**: 1347-1354.

5. Higuchi, T.; Ito, Y and Kawamura, I. (1967). *p*-Hydroxyphenylpropane components of grass lignin and the role of tyrosine-ammonia lyase in its formation. *Phytochemistry* **6**: 875-881.

6. Lapierre, C. (1986). Hétérogeneité des lignines de Peuplier:mise en evidence systématique. Thése de doctorat d'Etat. Université de Paris-Sud.

7. Lapierre, C.; Rolando, C and Monties, B. (1983). Characterization of poplar lignins acidolysis products: capillary gas-liquid and liquid chromatography of monomeric products. *Holzforschung* **37**: 189-198.

8. Lapierre, C.; Monties, B. and Rolando, C. (1985). Thioacidolysis of lignin: comparison with acidolysis. *Journal of Wood Chemistry and Technology* **5**: 277-292.

9. Lapierre, C., Monties, B. and Rolando, C. (1986a). Preparative thioacidolysis of spruce lignin: isolation and identification of main monomeric products. *Holzforschung* **40**: 47-50.

10. Lapierre, C.; Monties, B. and Rolando, C. (1986b). Thioacidolysis of poplar lignins: identification of monomeric syringyl products and characterization of guaiacyl-syringyl lignin fractions. *Holzforschung* **40**: 113-118.

11. Lapierre, C., Monties, B. and Rolando, C. (1987). Degradation of various lignins and lignin model compounds by thioacidolysis. *International Symposium on Wood and Pulping Chemistry, Paris, Proceedings of papers* **2**: 431-435.

12. Lapierre, C. and Rolando, C. (1988). Thioacidolyses of premethylated lignin samples from pine compression and poplar woods. *Holzforchung* **42**: 1-4.

13. Lundquist, K. (1976). Low-molecular weight lignin hydrolysis products. *Applied Polymer Symposium* **28**: 1393-1407.

14. Morrison, I.M. (1974). Structural investigations on the lignin-carbohydrate complexes of *Lolium perenne*. *Biochemical Journal* **139:** 197-204.

15. Nimz, H.H.; Robert, D.; Faix, O. and Nemr, M. (1981). Carbon-13 NMR spectra of lignins, 8. Structural differences between lignins of hardwood, softwoods, grasses and compression wood. *Holzforschung,* **35:** 16-26.

16. Reeves, J.B. (1987). Lignin and fiber compositional changes in forages over a growing season and their effects on *in vitro* digestibility. *Journal of Dairy Science* **70:** 1583-1594.

17. Scalbert, A. (1984). Caractérisation des lignines de paille de blé: fractionnement, associations avec les oses et les acides phénoliques. Thése de Docteur-Ingénieur. Institut National Agronomique de Paris-Grignon.

18. Scalbert, A.; Monties, B.; Guittet, E. and Lallemand, J.Y. (1986). Comparison of wheat straw lignin preparations I. Chemical and spectroscopic characterization. *Holzforschung* **40:** 11?-127.

19. Shimada, M.; Fukuzuka, T. and Higuchi, T. (1971). Ester linkages of *p*-coumaric acid in bamboo and grass lignin. *TAPPI* **54:** 72-78.

20. Timell, T.E. (1981). Recent progress in the chemistry ultrastructure and formation of compression wood. *International Symposium on Wood and Pulping Chemistry, Stockholm, Proceedings of Papers* **1:** 99-147.

21. Van Soest, P.J. (1964). New chemical procedures for evaluating forages. *Journal of Animal Science* **23:** 838-845.

22. Wacek. A.V. and Kratzl, K. (1947). Modellversuche zum Ligninproblem. *Osterreichische Chemie Zeitschriff* **48:** 36-40.

DETERMINATION OF PHENOLIC COMPOUNDS FROM LIGNOCELLULOSIC BY-PRODUCTS BY HPLC WITH ELECTROCHEMICAL DETECTOR

G.C. GALLETTI[1], R. PICCAGLIA[1], V. CONCIALINI[2], and M.T. LIPPOLIS[2]

[1]*Centro di Studio per la Conservazione dei Foraggi, C.N.R.-Istituto di Agronomia Generale e Coltivazioni Erbacee, Università di Bologna, Via Filippo Re 8, 40126 Bologna, Italy.*

[2]*Dipartimento di Chimica "G. Ciamician", Università di Bologna, Via Selmi 2, 40126 Bologna, Italy.*

SUMMARY

HPLC with electrochemical detector was applied to the determination of phenolic monomers in wheat straw, corn stalks, alfalfa stems and red oak leaves. Phenolic monomers were obtained from nitrobenzene oxidative hydrolysis of the native lignocellulose and lignin preparations. The diluted and acidified reaction mixtures were injected directly into HPLC avoiding solvent partitioning. Data are presented on detector reproducibility and recoveries from phenolic standard solution subjected to the classical solvent partitioning procedures.

INTRODUCTION

The characterization of phenolic compounds in lignocellulosic wastes is particularly important for the research aimed at finding methods for the economic utilization of agricultural surpluses. Both simple phenolics and complex macromolecules, such as tannins and lignins have attracted interest for various reasons: nutritionists relate the presence of these compounds to the poor quality of some forages[1] while others use the possible exploitation of such molecules as a relatively inexpensive source of chemicals.[2]

Native materials can be used with little or no manipulation for spectroscopic investigations such as infrared spectroscopy[3], near infrared spectroscopy[4], fourier transform infrared spectroscopy[5] and solid phase ^{13}C-nuclear magnetic resonance.[6] Other analytical methods require some degree of degradation of the phenolic macromolecules into smaller fragments suitable for determination by common chromatographic techniques.

Nitrobenzene oxidative hydrolysis is a well established technique for the cleavage of lignins into monomeric units.[7] Acidolysis[8] and, more recently, thioacidolysis[9] have also been proposed as degradative pathways for the study of lignin.

The separation of lignin-related phenolic monomers has been accomplished by Gas Chromatography (GC) after chemical derivatization[10] and by High Performance Liquid Chromatography (HPLC) using an ultraviolet detector (UV)[11]. Generally, HPLC requires less manipulations than G.C., since no chemical derivatization is necessary and water extracts can be injected although a G.C. method for the analysis of underivatized phenolic monomers has been published recently[12].

HPLC with an electrochemical detector (ElCD) can improve the determinations of phenolic compounds because of the better selectivity of ElCD compared to UV[13]. The analysis of phenolic monomers from nitrobenzene oxidative hydrolyzates is simplified by HPLC/ElCD since no purification of the reaction mixture by solvent partitioning is necessary prior to chromatographic injection[14,15]. Preliminary experiments on the evaluation of HPLC/ElCD for routine analysis of phenolic fractions in wheat straw have been reported previously[16]. Further results are presented here on the application of HPLC/ElCD to the quantitative determination of aromatic monomers from four lignocellulosic materials (wheat straw, corn stalks, alfalfa stems and red oak leaves) and their lignins prepared by H_2SO_4 extraction. The reproducibility of the detector is tested and data regarding the recovery of phenolics from the solvent partitionings[15] involved in the classical nitrobenzene procedure reported.

MATERIALS AND METHODS

Sample preparations

Lignocellulosic by-products, (wheat straw, corn stalks, alfalfa stems and red oak leaves) were collected at the USDA in Beltsville, MD, USA. Sample were homogenized in a UDY cyclone mill to 40 mesh. Lignins were prepared from each sample according to the Goering and Van Soest[17] procedure by treatment with 72% H_2SO_4 for three hours.

Alkaline nitrobenzene oxidation

Samples (100 mg) of lignocellulosics (50 mg of lignins) were added with 5 ml of 2M NaOH and 0.1 ml of nitrobenzene. The mixture was heated in a thick-glass tube with a Schott screw cap at 160°C for 2 hours in an oil bath under magnetic stirring. After cooling the mixture was diluted to 25 ml with water and filtered.

HPLC analysis

Filtrate (2 ml) were acidified to pH 2-3 with 1M HCl, diluted to 25 ml filtered through a 0.22 μm cartridge filter (Millipore, USA) and injected into HPLC (direct analysis).

Recovery test

Solutions (2 ml) of phenolic compounds at different concentrations (A: 0.1 g/l; B: 0.05 g/l; C: 0.01 g/l of each phenolic) in 2M NaOH were extracted with CH_2Cl_2 (3 x 2 ml) discarding the organic phase. The aqueous residue was acidified with 1 ml of 1M HCl and re-extracted with CH_2Cl_2 (4 x 2 ml). The combined organic phases were dried over anhydrous Na_2SO_4 and the solvent was evaporated in a rotary evaporator under vacuum. The residue was dissolved in methanol:water 1:5 (30 ml) and injected into HPLC (classical analysis). The waste aqueous and organic phases (2 and 3 respectively in Table 2) were saved to determine their possible content of phenolics.

Apparatus

A Rheodyne 7010 sample injector (0.02 ml loop), a M45 pump (Waters Ass., Milford, MA, USA) and an ESA Coulochem model 5100A electrochemical detector equipped with an analytical cell model 5011 set at 0.80 V were used for all chromatographic determinations. A reversed phase column (120 x 4.6 mm) Viospher C6, 5μm (Violet, Rome, Italy) was operated at room temperature with methanol:0.1% perchloric acid in water 15:85 as mobile phase at 1.00 ml/min. All solutions were filtered through a 0.22 μm cartridge filter (Millipore, USA) prior to HPLC use. An external standard was used for quantitation. All data were the average of two chromatographic analyses.

RESULTS AND DISCUSSION

The reproducibility of the HPLC/ElCD system was tested by replicate injections of standard phenolics during a three month period. Standard deviations of the factors response for 0.5-1 x 10^{-5} M phenolic solutions ranged from 10 to 13% (Table 1), suggesting that performance of the column and ElCD are relatively unaffected by a prolonged period of continuous analyses.

TABLE 1. Average HPLC/ElCD response factors (f_r) of standard phenolics (seven replicate injections during a three month period

	Compound	f_r	SD	%SD
1.	p-Hydroxyphenylacetic acid	0.223	0.022	9.9
2.	o-Hydroxyphenylacetic acid	0.348	0.040	11.5
3.	p-Hydroxybenzoic acid	0.345	0.037	10.5
4.	p-Hydroxybenzaldehyde	0.445	0.057	12.8
5.	Vanillic acid	0.981	0.122	12.4
6.	Vanillin	0.892	0.092	10.3
7.	Syringic acid	1.749	0.178	10.2
8.	p-Coumaric acid	0.651	0.085	13.1
9.	Syringaldehyde	1.802	0.184	10.2
10.	Ferulic acid	2.647	0.355	13.4

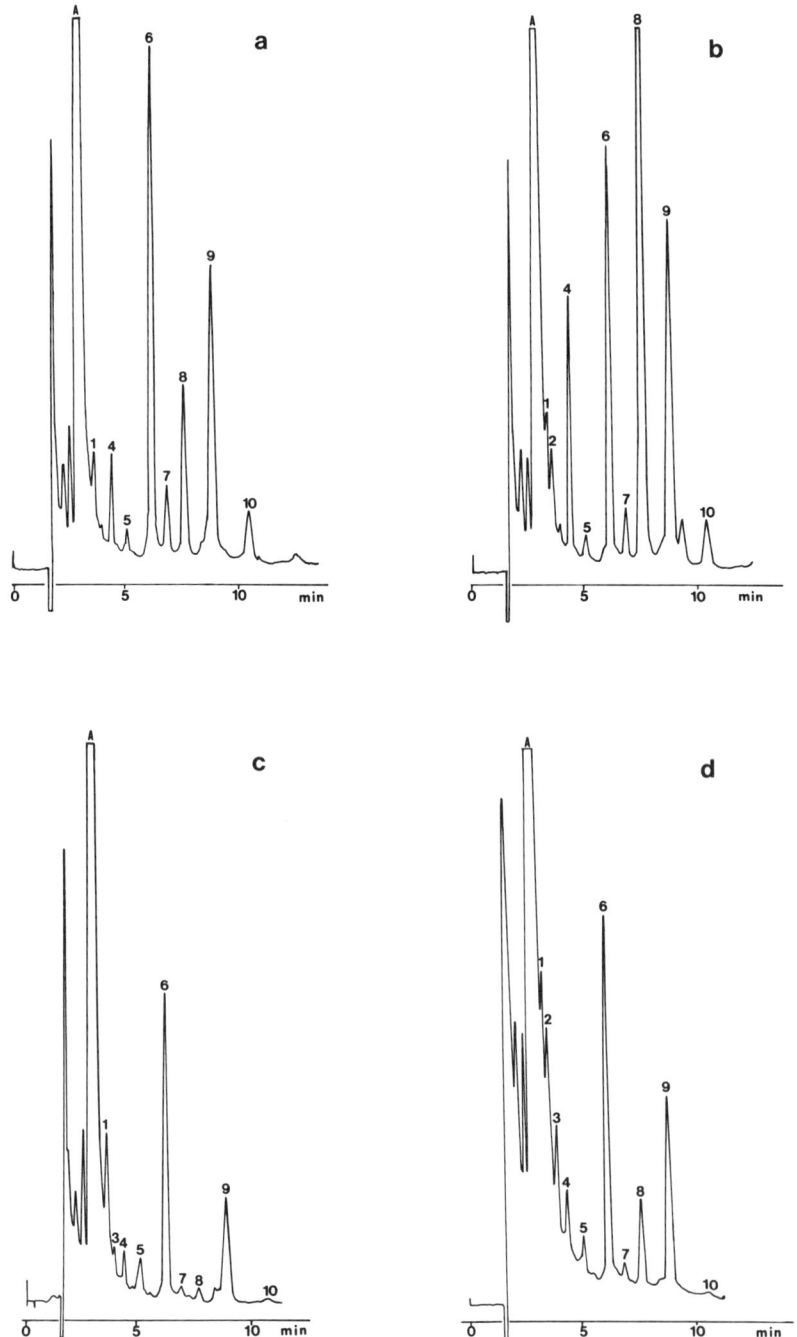

Figure 1. HPLC chromatograms of phenolic compounds from: a) wheat straw; b) corn stalks; c) alfalfa stems; d) red oak leaves. A, aniline. Peak numbers as in Table 1.

Previous workers[14,15] showed that more accurate results, compared to HPLC/UV, could be obtained by HPLC/ElCD direct analysis of nitrobenzene oxidative hydrolysates. ElCD is insensitive to the interference of non-electrochemically active UV absorbing substances (such as nitrobenzene), avoiding the need for solvent partitioning and any consequent loss of phenolics. Phenolic recoveries after solvent partitioning[15] (Table 2) were estimated to be generally dependent on the concentration of the solution, ranging from 32 to 37% for p-coumaric acid, 51-64% for vanillic acid, 53-75% for syringic acid and 57-82% for ferulic acid. The relatively higher recoveries of vanillin and syringaldehyde (45-80% and 57-97% respectively) may enhance the importance of these compounds in nitrobenzene oxidative hydrolysates, when solvent partitioning is adopted.

TABLE 2. Distribution of phenolics among the phases of the nitrobenzene classical analytical procedure. A: 0.1 g/l of each phenolic; B: 0.05 g/l; C: 0.01 g/l. 1: final CH_2Cl_2 phase; 2: waste H_2O; 3: waste CH_2Cl_2. ElCD measurements expressed as % of the direct results

Compound	A			B			C		
	1	2	3	1	2	3	1	2	3
Vanillic acid	50.96	39.00	4.11	76.26	19.52	3.88	64.34	26.15	3.80
Vanillin	45.36	41.09	4.88	79.78	4.55	3.74	58.47	2.56	4.29
Syringic acid	53.09	38.11	2.09	74.95	9.74	3.67	75.35	12.49	3.44
p-Coumaric acid	36.99	38.52	14.41	34.74	47.80	2.66	31.78	42.12	3.34
Syringaldehyde	56.98	38.62	-	96.95	1.74	3.34	89.51	-	2.72
Ferulic acid	56.54	41.92	-	82.35	2.06	2.65	70.77	-	-

The phenolic composition of the four lignocellulosic materials is summarised in Table 3, and the corresponding chromatographic profiles are shown in Figure 1. Corn stalks showed the higher content of p-hydroxybenzaldehyde, syringaldehyde and, particularly, p-coumaric acid (2.10% versus 0.45, 0.23 and 0.03 for wheat straw, red oak leaves and alfalfa stems, respectively). Vanillin, syringic acid and ferulic acid were more abundant in corn stalks than in alfalfa stems and red oak leaves. Corn stalks showed the highest total phenolic content. The highest percentages of vanillin, syringic acid and ferulic acid were found in wheat straw, whereas alfalfa stems and red oak leaves were characterized by a lower total phenolic content. p-Hydroxybenzaldehyde, syringic acid and ferulic acid contents were particularly low in alfalfa stems and red oak leaves. The p-coumaric acid content was 0.03% in alfalfa stems. p-Coumaric acid, ferulic acid and syringic acid showed the widest concentration changes among the different materials, although these compounds were not the most abundant.

The phenolic composition of lignins derived from the lignocellulosics is shown in Table 4. Figure 2 shows the corresponding chromatographic patterns. Vanillin and syringaldehyde were the most abundant compounds. Corn stalk lignin showed the highest p-hydroxybenzaldehyde, syringaldehyde and p-coumaric acid content. Compared to the original materials, lignins yielded

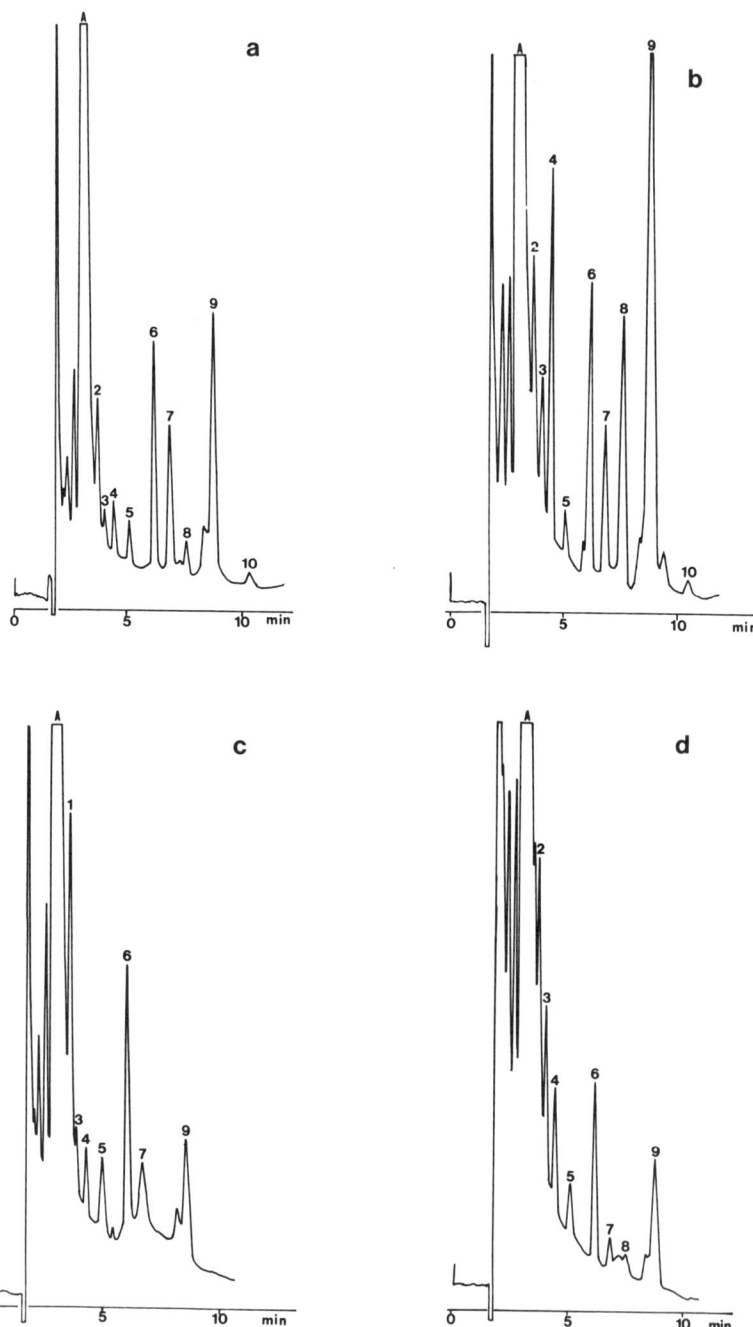

Figure 2. HPLC chromatograms of phenolic compounds from H₂SO₄ lignin preparations from: a) wheat straw; b) corn stalks; c) alfalfa stems; d) red oak leaves. A, aniline. Peak numbers as in Table 1.

TABLE 3. Phenolic composition of lignocellulosic by-products (% of the original material)

	Compounds	Wheat straw	Corn stalk	Alfalfa stem	Red oak leaf
1.	*p*-Hydroxyphenylacetic acid	0.05	0.03	0.10	0.08
2.	*o*-Hydroxyphenylacetic acid	-	0.11	-	0.14
3.	*p*-Hydroxybenzoic acid	-	-	0.03	0.15
4.	*p*-Hydroxybenzaldehyde	0.19	0.52	0.07	0.11
5.	Vanillic acid	0.12	0.10	0.16	0.15
6.	Vanillin	2.14	1.73	1.26	1.52
7.	Syringic acid	0.55	0.40	0.07	0.16
8.	*p*-Coumaric acid	0.45	2.10	0.03	0.23
9.	Syringaldehyde	2.50	2.84	0.09	1.66
10.	Ferulic acid	0.56	0.49	0.03	0.05

relatively higher percentages of vanillic acid and syringic acid than vanillin and syringaldehyde. Low amounts of *p*-coumaric acid and ferulic acid were found in lignins. The total phenolic content showed the same pattern as in the original materials (Fig. 3). When compared to native materials, red oak leaves appear to yield relatively higher percentages of insoluble substances after H_2SO_4 extraction, probably due to the high tannin content[18] capable of undergoing extensive condensation during the acid treatment.

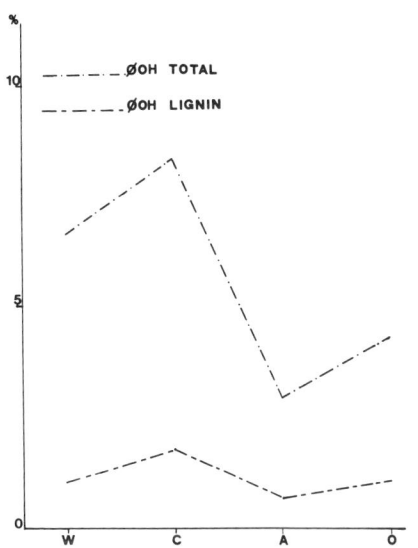

Figure 3. Yield of phenolic compounds from native lignocellulosics and lignin preparations: W, wheat straw; C, corn stalks; A, alfalfa stems; O, red oak leaves.

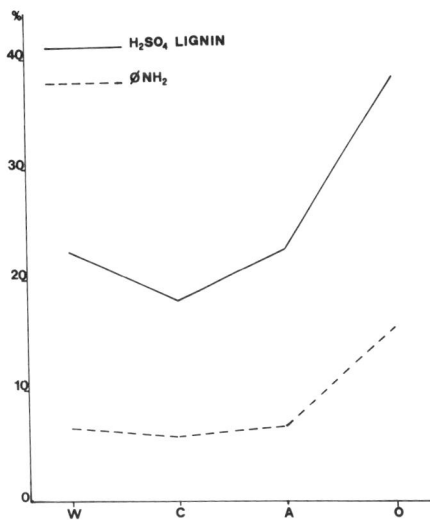

Figure 4. Yield of lignin and aniline from native lignocellulosics: W, wheat straw; C, corn stalks; A, alfalfa stems; O, red oak leaves.

Table 4. Phenolic composition of lignins prepared from lignocellulosic by-products (% of the original material)

	Compound	Wheat straw	Corn stalk	Alfalfa stem	Red oak leaf
1.	p-Hydroxyphenylacetic acid	-	0.03	0.06	-
2.	o-Hydroxyphenylacetic acid	-	-	-	0.12
3.	p-Hydroxybenzoic acid	-	-	0.01	0.09
4.	p-Hydroxybenzaldehyde	0.02	0.12	0.03	0.09
5.	Vanillic acid	0.04	0.03	0.07	0.09
6.	Vanillin	0.20	0.20	0.22	0.28
7.	Syringic acid	0.23	0.19	0.06	0.09
8.	p-Coumaric acid	0.02	0.11	-	0.02
9.	Syringaldehyde	0.48	0.83	0.21	0.40
10.	Ferulic acid	0.03	0.03	-	-

Differences in the phenolic composition of the studied materials could be shown by HPLC/EICD, particularly with respect to p-coumaric acid and ferulic acid. Aniline produced from the reduction of nitrobenzene could be quantitated (Figure 4). The resemblance of aniline and lignin yields may suggest further research in order to ascertain the role of aniline as an oxidatability index of the lignocellulosic matrices.

REFERENCES

1. Jung, H.G. and Fahey, G.C. Jr. (1983). Nutritional implications of phenolic monomers and lignin: a review. *Journal of Animal Science* **57**: 206-219.

2. Sachetto, J.-P.; Armanet, J.-M.; Roman, A. and Johansson, A. (1987). The fractionation of lignocellulosics for the production of chemicals. In *Degradation of lignocellulosics in ruminants and in industrial processes* Commission of the European Communities. Van der Meer, J.M., Rijkens B.A., Ferranti, M.P. (Eds.) Elsevier Applied Science, London, New York p. 89-96.

3. Hergert, H.L. (1971). Infrared spectra. In: *Lignins occurrance, formation structure and reactions*. Sarkanen, K.V. and Ludwig, C.H. (Eds.) Wiley interscience, New York, London, Sidney, Toronto, p. 267-297.

4. Reeves, J.B. III (1988). Near infrared spectroscopic analysis of lignin components in sodium chlorite-treated and untreated forages and forage by-products. *Journal of Dairy Science* **71**: 388-397.

5. Buta, J.G. and Galletti, G.C. (accepted for publication). FT-IR investigation of lignin components in various agricultural lignocellulosic by-products. *Journal of the Science of Food and Agriculture*.

6. Himmelsbach, D.S. and Barton, F.E. II (1980). ^{13}C-nuclear magnetic resonance of grass lignins. *Journal of Agricultural and Food Chemistry* **28**: 1203-1208.

7. Freudenberg, K.; Lautsch, W. and Engler, K. (1940). Die Bildung von Vanillin aus Fichtenlignin. *Berichte der Deutschen Chemischen Gesellschaft* **73**: 167-171.

8. Lundquist, K. (1976). Low molecular weight lignin hydrolysis products. *Applied Polymer Science* **28**: 1393-1407.

9. Lapierre, C.; Monties, B. and Rolando, C. (1988). Thioacidolyses of lignin: comparison with acidolysis. *Journal of Wood Chemistry and Technology* **5**: 277-292.

10. Cymbaluk, N.F. and Neudoerffer, T.S. (1970). A quantitative gas-liquid chromatographic determination of aromatic aldehydes and acids from nitrobenzene oxidation of lignin. *Journal of Chromatography* **51**: 167-174.

11. Hartley, R.D. and Buchan, H. (1979). High performance liquid chromatography of phenolic acids and aldehydes derived from plants or from the decomposition of organic matter in soil. *Journal of Chromatography* **180**: 139-143.

12. Fritz, J.O. and Moore, K.J. (1987). Separation and quantification of lignin derived phenolic monomers using high-resolution gas chromatography. *Journal of Agricultural and Food Chemistry* **35**: 710-713.

13. Chiavari, G.; Vitali, P. and Galletti, G.C. (1987). Electrochemical detection in the high-performance liquid chromatography of polyphenols (vegetable tannins). *Journal of Chromatography* **392**: 427-434.

14. Chiavari, G.; Concialini, B. and Galletti, G.C. (1988). Electrochemical detection in the high-performance liquid chromatographic analysis of plant phenolics. *Analyst* **113**: 91-94.

15. Galletti, G.C.; Piccaglia, R.; Chiavari, G. and Concialini, G. (submitted). Quantitative analysis of phenolics from wheat straw using HPLC with electrochemical detector. *Journal of Agricultural and Food Chemistry*.

16. Galletti, G.C.; Piccaglia, R.; Chiavari, G. and Concialini, V. (in press). Analysis of lignin in straw by HPLC with electrochemical detector. In: *Methods of straw evaluation in ruminant feeding*. Commission of the European Communities.

17. Goering, H.K. and Van Soest, P.J. *Forage fiber analyses*. Agricultural Handbook n°. 379. Agricultural Research Service, United States Department of Agriculture, Washington, D.C.

18. Rice, E.L. (1984). *Allelopathy*. Academic Press, Orlando, San Diego, San Francisco, New York, London, Montreal, Sidney, Tokio, Sao Paulo, p.85.

ELECTROCHEMICAL APPROACHES TO THE OXIDATIVE DEGRADATION OF LIGNINS AND LIGNOCELLULOSIC MATERIALS

V. CONCIALINI[1], M.T. LIPPOLIS[1], G.C. GALLETTI[2] and R. PICCAGLIA

[1]Dipartimento di Chimica "G. Ciamician", Università di Bologna Via Selmi 2, 4012(Bologna, Italy

[2]Centro di Studio per la Conservazione dei Foraggi, CNR-Istituto di Agronomia General(e Coltivazioni Erbacee, Università di Bologna, Via Filippo Re 8, 40126 Bologna, Italy

SUMMARY

Indirect electro-oxidation of a sample of corn stalks in alkaline solution using potassium ferrocyanide as redox catalyst was performed. The concentration of the major phenolic compounds (p-coumaric and ferulic acids) as a function of charge passed through the solution, was followed by HPLC. The consecutive oxidations of p-coumaric and ferulic acids to the corresponding aldehydes and carboxylic acids were demonstrated.

INTRODUCTION

Technical lignins derived from the industrial processing of wood an annual plants such as straw and other lignocellulosic waste are of great potentia interest as renewable sources of aromatic materials.

Oxidative chemical degradation has been extensively studied as a route t utilization and various oxidants have been proposed. The alkaline nitrobenzen oxidation[1,2] is the oldest and best known method and its mechanism of actio has been extensively studied[3,4]. Ceric ammonium nitrate also has been used i the oxidative-cleavage of some lignin model compounds[5]. Chemical oxidation often require hazardous and expensive reagents and in the case of industri processes serious environmental problems connected with spent reagents ma arise.

Electrochemical processes are a cleaner and more controlled method c oxidation. Electroorganic syntheses are being increasingly utilized by organi chemists[6]. Excellent reviews on electrolytic oxidations of lignins are given b Chum and Osteryoung[7], Chum and Baizer[8] and Utley.[9]

Limosin et al.[10], in the course of an extensive study on several dimeri model compounds of lignin, concluded that interesting results on th electrooxidation of lignin were complicated by the formation of a polymeri non-conductive film on the Pt anode, a problem already reported in the cycli voltammetry of phenolic compounds[11]. They also reported that the oxidatio of some lignins on Pt could be achieved with the addition of $K_4Fe(CN)_6$ as a electron carrier, thereby limiting the inhibition of the anode[12].   The use of potassium ferrocyanide is proposed also in a U.S. Patent[13] as a mediator for the delignification of lignocellulosic material.

Very recently two papers by Yoshiyama *et al.* reported the anodic degradation of hardwood[14] and softwood[15] lignins. The tremendous amount of research carried out by Soviet authors in this field is quoted by Yoshiyama in his extensive bibliography and the possibility of obtaining useful chemical compounds from electrochemical treatments of lignins is emphasised.   The electrolysis was performed galvanostatically in alkaline solutions at 50°C and the degradation products analyzed by HPLC. The yield and the distribution of degradation products were greatly affected by the current density. In fact, under galvanostatic conditions, the potential of the working electrode is not controlled so that further oxidations may take place at more positive anodic potentials, with a consequent loss of the desired products.   This point will be further mentioned in the following discussion.

As already suggested by Limosin *et al.*[12] the use of a redox catalyst or an electrochemical mediator permits the oxidation of lignins in bulk solution by ferricyanide ions with the formation of ferrocyanide ions which regenerate anodically in initial oxidizing ion. Such an indirect oxidation has recently found increasing interest and applications[16-18].

The following scheme depicts the cycles involved in this technique.

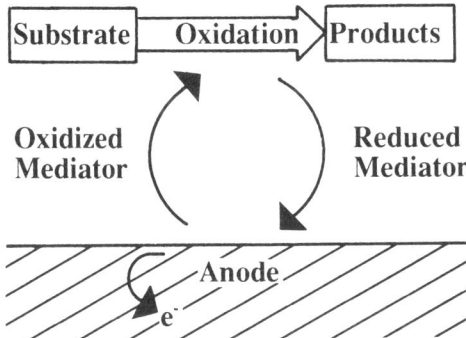

This system seems particularly suitable for substances such as lignins or lignocellulosic materials, since the "chemical oxidations" of the substrate are possible even it it is not completely dissolved or it is a suspension or emulsified. The "direct electroxidation" would not proceed sufficiently fast due to the extremely slow electrode reactions. The mediator acts as a catalyst in the electron transfer from the electrode to the substrate so that a relatively high current density may be realized without increasing too much the electrode potential.   A large number of redox catalysts both organic and inorganic are available and the choice of mediator is obviously critical.

In this preliminary work we re-examine the redox system Fe^{III}/Fe^{II}, coupling the electrochemical measurements with HPLC analysis of the reaction mixture.

MATERIALS AND METHODS

Corn stalks were collected at the USDA in Beltsville, MD, U.S.A. and were homogenized in an UDY cyclone mill to 40 mesh.

The electrochemical equipment consisted of a Potentiostat (AMEL Model 552) and a digital integrator (AMEL Model 721). Voltammetric curves and electrolysis currents were recorded on an X-Y recorder (HP Model 7040A).

A three electrode thermostatted cell was employed with saturated calomel electrode (SCE) as reference, a Pt coil, separated from the solution by a fritted disk, as counter electrode and either a Pt electrode (AMEL Model 492), with periodical renewal of the diffusion layer or a Pt foil for the electrolysis as working electrodes.

HPLC determinations were performed using a Rheodyne 7010 sample injector (0.01 ml loop), a Varian 2510 pump with a Varian 2550 variable detector set at 280 nm and a Varian 4290 Integrator. A reversed phase column Erbas' C18 (10μm) was used with a mobile phase of 30:70 of methanol: 0.1% perchloric acid in water at a flow rate of 1.0 ml/min.

RESULTS AND DISCUSSION

Voltammetric behaviour of the ferrocyanide-ferricyanide system in the presence of lignin is reported in Figure 1. Graph (a) shows the cyclic voltammogram for the oxidation of ferrocyanide in 2M NaOH, typical of a one electron reversible redox couple, while Graph (b) shows the corresponding voltammogram in the presence of corn stalks. In contrast to (a) the anodic peak is increased, indicating the reaction of the ferricyanide ion with the substrate regenerating ferrocyanide.

A controlled potential electrolysis (CPE) was carried out at a potential of +0.3V vs. SCE, recording the current (initially of the order of 100 mA) and following the relative concentrations of Fe^{II} and Fe^{III} voltametrically utilizing the Pt electrode which was mechanically pulsed to renew the diffusion layer. The solution contained 50 mg of milled corn stalks in 20 ml of 2M NaOH, in the presence of 0.02 M $K_4Fe(CN)_6$.

Typically 50 to 100 Coulombs were passed through the solution in addition to those calculated for the monoelectronic process Fe^{II} -> Fe^{III} so that consumption of 1-2 Coulombs/mg substrate can be evaluated. During the CPE the cell solution was periodically analyzed by HPLC.

The most abundant of the phenolic products found in the corn stalks are p-coumaric and ferulic acids deriving from p-hydroxyphenyl and guaiacyl lignin units respectively.

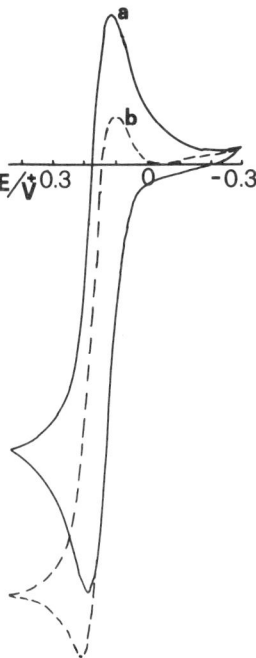

Figure 1. a) Cyclic voltammogram of 0.02 M K$_4$[Fe(CN)$_6$] in 2 M NaOH. Sweep rate 20 mV s^{-1}, temp. 50°C. b) as a) after the addition of 2.5 mg/ml of milled corn stalks.

In Figure 2 the dependence of the concentration of the *p*-coumaric acid on time and charge is reported, together with the appearance of new peaks assigned to *p*-hydroxybenzaldehyde and *p*-hydroxybenzoic acid. A close resemblance of these curves with those found in the kinetics of consecutive reactions is apparent. The concentration of *p*-coumaric acid, initially present as the major component (about 1% of the original material), decreased with time and charge, while that of *p*-hydroxybenzaldehyde increased passing through a maximum, then decreased practically to zero, and *p*-hydroxybenzoic acid, initially absent, increased reaching a constant value at the end of the electrolysis.

Ferulic acid and its oxidation products, vanillin and vanillic acid, also showed a similar behaviour, but starting from a smaller initial concentration (about 0.1%).

The same experiments performed at a higher temperature (75°C) show exactly the same trend in the concentrations of the above phenolic compounds. However in this case the time scale was shortened so that in about 1 hour all the *p*-coumaric and ferulic acids were oxidized, with the consumption of the same charge (ca. 1.5 Coulombs/mg substrate).

During the electrolysis a few peaks, with short retention times, increased in size, due to the possible formation of unidentified polyhydroxylated monomers.

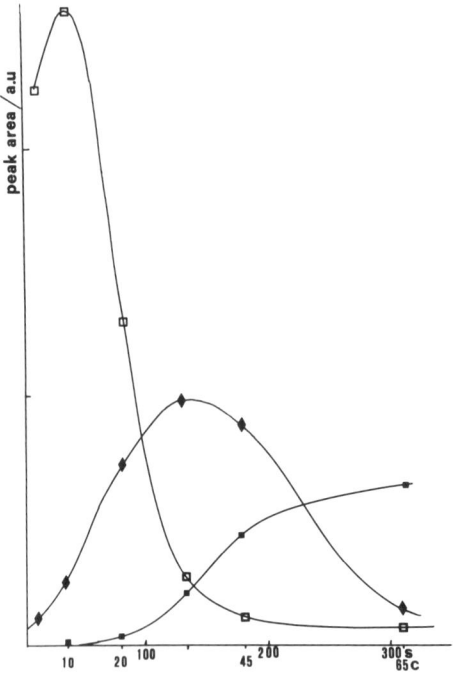

Figure 2. Dependence of the peak areas of □ p-coumaric acid, ♦ p-hydroxybenzaldehyd͏e and ■ p-hydroxybenzoic acid on time (seconds) and charge (coulombs, corrected for th͏e oxidation of ferrocyanide). Solution: 2M NaOH, 0.02 M $K_4[Fe(CN)_6]$, 2.5 mg/ml mille͏t corn stalks. Temp. 50°C.

From the preceding results it appears that control of the degradativ͏e procedure in the oxidative cleavage of lignin is important. Since some of th͏e products may be formed and consumed during the oxidation procedur͏e ambiguous results may be obtained concerning structural information. Thi͏s difficulty was pointed out by Lapierre et al.[19] in proposing their thioacidolysi͏s procedure which preserves the phenylpropane skeleton during the cleavage ͏of alkylaryl ether bonds.

In the electrochemical oxidation processes it is therefore important t͏o control the electrode potential. In the case of mediated electroxidatio͏n differences in the standard potential of the redox couple and the reactio͏n conditions may lead to a diversity of products. Therefore work is currently i͏n progress to study the dependance of the anodic degradation of various ligni͏n and lignocellulosic by-products on the redox properties of the mediators.

REFERENCES

1. Freudenberg, K.; Lautsch, W. and Engler, K. (1940). Die Bildung vo͏n Vanillin aus Fichtenlignin. *Chemische Berichte* **73**: 167-171.

2. Creighton, R.H.J.; McCarthy, J.L. and Hibbert, H.J. (1941). "Aromatic aldehydes from Spruce and Maple woods". *Journal of the American Chemical Society* **63:** 312.

3. Chang, H.M. and Allan, G.G. (1971). In: "Oxidation" *Lignins. Occurrence, formation, structure and reactions* (Sarkanen, K.V. and Ludwig, C.H. Eds.), Wiley-Interscience: New York. pp. 433-485.

4. Schultz, T.P. and Templeton, M.C. (1986). Proposed mechanisms for the nitrobenzene oxidation of lignin. *Holzforschung* **40:** 93-97.

5. Fischer, H.T., Dershem, M.S. and Schultz, T.P. (1988). "Ceric ammonium nitrate oxidative cleavage reaction of some lignin model compounds: role of the benzylic hydroxil group". *Journal of Organic Chemistry* **53:** 1504-1507.

6. Torii, S. (1985). *Electroorganic Synthesis*, Monographs in modern chemistry, Part I, Oxidations. VCH Publishers Inc. Weinheim, Germany.

7. Chum, H.L. and Osteryoung, R.A. (1982). Survey of the Electrochemistry of some biomass-derived compounds - Solar Energy Research Institute for the U.S. Department of Energy. National Technical Information Service U.S. Department of Commerce 5285 Port Royal Road Springfield, VA 22161.

8. Chum, H.L. and Baizer, M.M. (1985). *Electrochemistry of Bio-mass and Derived Materials*. American Chemical Society, Washington D.C.

9. Utley, J.H. (1985). The electrochemical conversion of bio-mass-derived compounds. In: *Fundamentals of Thermochemical Bio-mass Conversion*. (R. Overend, T. Milne and L. Mudge Eds.) Elsevier Applied Science, London, U.K.

10. Limosin, D.; Pierre, G.; Cauquis, G.; Chopin, J.; Chadenson, M. and Hauteville, M. (1985). Etude Electrochimique de quelques Composés Diméres Modéles de la Lignine. *Holzforschung* **39:** 91-98.

11. Chiavari, G.; Concialini, V. and Galletti, G.C. (1988). Electrochemical detection in the high-performance liquid chromatographic analysis of plant phenolics. *Analyst* **113:** 91-94.

12. Limosin, D.; Pierre, G. and Cauquis, G. (1986). Oxidation electrochimique de quelques echantillons de lignine en solution aqueuse basique. *Holzforschung* **40:** 31-36.

13. Godsay, M.P.; Hull, M.N. and Yasnovsky, V.M. (1986). Process for the delignification of lignocellulosic material with oxygen, ferricyanide and a protector. U.S. Patent 4,622,100.

14. Yoshiyama, A.; Nonaka, T.; Chou, T.C. and Tien, H.J. (1988). Anodic degradation of lignins I. Softwood lignins. *Mozukai Gakkaishi* **34:** 155-161.

15 Yoshiyama, A.; Nonaka, T.; Chou, T.C. and Tien, H.J. (1988). Anodic degradation of lignins II. Softwood lignins. *Mozukai Gakkaishi* **34:** 275-280.

16. Simonet, J. (1983). Electrogenerated reagents. In: *Organic Electro-chemistry, an Introduction and a Guide,* 2nd ed. (M.M. Baizer and H. Lung Eds.) M. Dekker, New York, pp. 843-871.

17. Steckhan, E. (1987). Organic syntheses with electrochemically regenerable redox systems. In: *Topics in Current Chemistry, Electrochemistry I* (E. Steckham, Ed.) Springer-Verlag, Berlin.

18. see Ref. (6). p.279. Indirect Electrooxidation and the Use of Electron Carriers (Mediators).

19. Lapierre, C.; Monties, B. and Rolando, C. (1988). Thioacidolysis of lignin: comparison with acidolysis. *Journal of Wood Chemistry and Technology* **5:** 277-292.

METHODS FOR THE EVALUATION OF LIGNIN PROPERTIES SUITABLE FOR CONVERSION

A. HÜTTERMANN, O. MILSTEIN, B. NICKLAS, J. TROJANOWSKI,
A. HAARS, AND A. KHARAZIPOUR

Forstbotanisches Institut der Universität Göttingen, Göttingen, W. Germany

SUMMARY

Lignin, a natural polymer composed of C6-C3-phenolics, in theory at least, seems to be a very attractive feedstock with many potential uses for the polymer industry. However the market potential of lignin is greatly limited at present because lignins derived from the pulp and paper industry have rather unattractive properties, notably i) a low content of hydroxyl groups, ii) unsuitable molecular weight, which is either too high (lignosulphonate and Kraft lignin) or too low (organosolv lignin) and iii) the wrong polarity, being again either too high (lignosulphonate) or too low (Kraft and organosolv lignin).

A strategy for enzymic modification of lignin for technical use should therefore concentrate on achieving a homogenous, relatively pure lignin of reasonably high molecular weight and with a high reactivity provided by a high content of reactive (i.e. hydroxyl) groups. The enzymes described so far which are able to modify lignin are ligninase (lignin peroxidase), laccase and poly-blue oxidase. Although these enzymes have different reaction mechanisms they have a common requirement for phenolic hydroxyl groups for their reaction. Lignins from monocotyledons are characterised by a high content of phenolic groups and could form feedstocks for a polymer industry based on the use of fungal enzymes, provided they are available in high amounts and do not loose too much of their reactivity during the pulping process.

INTRODUCTION

The idea of using natural polyphenols (lignin) as feedstock for polymer production or as adhesives in wood composites is not new and has encouraged quite a number of scientific endeavours (for a recent review see ref.[1]). The basic rationale behind these studies being:

i) to find ways to replace expensive petrochemical resins by this comparatively cheap and renewable raw material;

ii) to construct polymers which are much more easily degraded by microorganisms after their disposal;

iii) to increase the overall benefit of the utilisation of renewable resources like timber or straw by obtaining additional credit for the utilisation of lignin. Any biomass conversion process which delivers lignin in a form which could be useful as raw material for other purposes at a price higher than its heating equivalent, would significantly change the overall economic balance. This could in many cases mean a switch from no or negative returns to positive ones.

For the design of lignin-based polymers, the following strategies are employed at present:

 i) derivatising the lignin molecule with small molecules like epichlorhydrins prior to incorporation into composite plastics;

 ii) forming graft polymers based on lignins;

 iii) extruding mixtures of lignin and another polymer.

All lignin available on the market today comes from the two main procedures currently employed for the pulping of wood to cellulose and paper the sulphite and the sulphate processes. Although lignin occurs in the plant cell walls as a compound with a molecular weight of about 10,000 and with many reactive groups in the molecule, the lignins available from the cellulose production are less attractive for further polymer production. The harsh reaction conditions produced during the technical pulping process result in the condensation of the lignins and a reduction in the number of functional groups. The properties of the commercially available lignins thus can be summarized as follows. They all have a low content of hydroxyl groups in the molecule and high molecular weights. Lignins from the sulphite process are extremely polar, owing to the sulphonate groups and sulphate esters in the molecule and are highly water soluble. In contrast Kraft lignins, derived from the sulphate process, have a very low polarity and are water insoluble. Both lignins therefore are rather unsuitable as feedstocks for polymer production. In addition, little hope can be drawn from the newly designed pulping processes like the Organosolv procedure. Lignin coming from this process is of very low reactivity and is also water insoluble. Its molecular weight is rather low and very probably still too small to be suitable for a prepolymer for plastics production. which needs a defined product of higher molecular weight (c.f. ref. [2]).

A strategy for enzymic modification of lignin for improvement of its technical use thus should concentrate on the following goals: to achieve a rather homogenous, relatively pure lignin with a high reactivity provided by a high content of reactive groups.

This communication reviews the present state of knowledge of lignin-modifying enzymes from lignin-degrading fungi which could be used in the conversion of lignins and reports on recent findings on the distribution of these enzymic activities among different species of fungi, coming from different ecological niches. In addition, we will present our work on the formulation of a two-component adhesive for particle boards composed of lignin and the enzyme laccase. The consequences of these findings with regard to lignin properties will be discussed.

EXTRACELLULAR ENZYMES FROM FUNGI WHICH CONVERT LIGNIN

The following enzyme activities able to modify the structure of native lignin have been characterised to date:

i) ligninase, (lignin peroxidase), veratryl peroxidase, the enzyme isolated and characterised by Kirk and his collaborators[3] (for updated reviews refs. [4,5]). The enzyme is considered to be the main lignolytic system in white-rot fungi and is measured by its ability to catalyse the oxidation of veratryl alcohol.

ii) laccase, (polyphenol oxidase, E.C.1.10.3.2). This enzyme activity is widely distributed and is reported to have quite different physiological functions[6]. It has been known for a long time to be able to polymerise phenols, its role in lignin degradation was first discussed by Ander and Eriksson[7] but this still remains unclear (for a recent review see ref[1])

iii) poly-blue oxidase. This enzymic activity was first described by Glenn and Gold[8]. It oxidises the lignin-model compound poly-blue, which is a polymeric dye.

All the three enzymatic activities mentioned above have the advantage that they can be determined rather easily by spectroscopic assays.

The most important gross changes in the lignin molecule which can be induced by enzymes and which are important for a possible industrial use, are:

  i) solubilisation of the lignin molecule;
  ii) demethylation;
  iii) changes in the phenolic and aliphatic hydroxyl groups;
  iv) changes in the molecular weight distribution.

In our present studies we screened for the first two of the above criteria in addition to the enzymic activities mentioned above. Data on changes in the molecular weight distribution have been given elsewhere[9,10]. Although work on lignin degradation has mainly focussed on one single, albeit rather suitable organism, the white-rot fungus *Phanaerochaete chrysosporium* (c.f.[5]), we decided to examine the lignin converting capacity of different fungi coming from different niches in the ecosystem in which fungi degrade wood.

The results of this screen are summarised in Table 1. The data clearly indicated that, within the variety of different species examined, no correlation exists between the activities of the different enzymes which were studied for their lignin modifying activity. The same holds true for the pattern of those activities during the growth cycle (data not shown). Also surprising was the poor correlation observed between the ability to solubilise organosolv lignin and the activity of the lignin peroxidase, although a significant correlation was found to exist between lignin solubilisation and poly-blue oxidase activity

TABLE 1. Peroxidase and oxidases production by selected wood-inhabiting fungi and their capacity to demethylate and to solubilise lignin

Fungus	Units x 10^3/mg mycelium			Demethylation[d] %	Solubilisation[e] %
	Lignin-peroxidase[a]	Poly-blue oxidase[b]	Laccase[cc]		
Trametes versicolour	1.6	20.8	2.9	29.5	52.1
Polyporus pinsitus	1.7	33.3	5.6	28.0	65.1
Phallus impudicus	2.5	14.0	0	27.1	13.3
Oudemansiella radicata	2.2	5.1	1.4	16.0	20.3
Bjerkandera adusta	7.6	8.6	0	16.8	19.5
Pleurotus florida F6	1.7	26.5	1.7	14.6	35.3
Pleurotus florida PP	1.7	20.6	3.9	14.0	45.3
Polyporus platensis	1.6	24.0	0.9	14.0	60.0
Ustulina deusta	1.0	5.6	0.8	14.0	19.2
Polyporus varius	3.2	7.1	0	12.0	16.9
Xylaria polymorpha	1.7	0.9	0.3	12.0	15.8
Phlebia radiata	3.9	25.0	4.7	11.0	56.8
Polyporus brumalis	1.5	23.4	1.8	10.6	56.1
Merulius tremellosus	2.8	27.0	1.2	8.0	58.9
Daedaeopsis confragosa	3.4	13.4	0	5.0	31.2

a   1 Unit = increase of A_{310}/min/1 ml medium using veratric alcohol and H_2O_2 at pH 3.
b   1 Unit = decrease of the adsorbance ratio A_{595}/A_{482}/ml medium after 24 h incubation using 0.01 % Poly Blue.
c   1 Unit = change of A_{645}/min/ml medium using 0.01 % TMB.
d   Figures indicate accumulative release of $^{14}CO_2$ using $^{14}CH_3O$-organosolv-lignin.
e   Figures indicate ^{14}C-water-solubles in culture as % of initial ^{14}C-activity using ^{14}C-β-organosolv-lignin.

(r=0.82). Therefore we considered this enzymic activity as a potentially interesting tool for solubilising lignin and part of our future work will concentrate on this enzymic activity.

LACCASE FROM WHITE-ROT FUNGI

Although there is no stringent correlation between laccase activities of fungi and their ability to degrade lignin (e.g. Table 1), a prominent role for this enzyme in lignin degradation has been proposed by several authors[4,7,11]. Several other reports have demonstrated that, in the presence of laccase, lignin is polymerised both *in vivo*[12,13] and *in vitro*[14,15]. Laccase acts on phenolics via a non-specific oxidation which generates quinoid intermediates. This results in the initial formation of dimers which subsequently spontaneously polymerise[16,17,18]. It is therefore reasonable to expect that the low molecular substances coming from enzymatically degraded lignin are readily repolymerised and polymerisation reactions may dominate over the depolymerisation processes during lignin transformation *in vitro*[19].

In addition to the polymerisation, which results in the crosslinking of lignin, this group of enzymes can catalyse another very important step for the activation of lignin molecules. They are able to hydroxylate phenolic substrates, thus introducing new phenolic hydroxyl groups into the lignin molecule and rendering more active sites on the molecule susceptible for subsequent reactions. These catalytic properties makes the enzyme a very interesting candidate for a variety of possible uses in lignin biotechnology.

LACCASE IN ORGANIC SOLVENTS

Lignin in its natural state is completely water insoluble and it is to be expected that any procedure which releases lignin from the lignocellulose complex under relatively mild conditions will also yield insoluble material. It therefore would be desirable to find out if and under what conditions at least some of the lignin-converting enzymes could act under conditions in which lignins are soluble. We therefore looked for means to retain the activities of the above mentioned enzymes in organic solvents, using laccase in the first instance.

It has recently been shown that a number of enzymes, when immobilized, can express their activity in a reaction medium in which the bulk of water has been replaced by organic solvents[20,21,22,23,24]. Furthermore, Klibanov and co-workers[25] have shown that lignin could be depolymerised with horseradish peroxidase in organic media, an observation, however, which still is under discussion[26].

For our studies we used laccase from the Basidiomycete *Trametes versicolor*, which was purified using DEAE-Sepharose A-50 chromatography. When a lyophilised powder of the purified laccase was added to a solution of the laccase substrates dimethoxyphenol or syringaldazine in water-saturated ethylacetate, the colour of the substrates rapidly turned to yellow or red-violet respectively, indicating that the usually observed laccase reaction takes place under these conditions.

However, the enzyme-reaction system described above has several disadvantages similar to those described for the horseradish peroxidase in organic solvents[24]. Since the enzyme does not dissolve in the organic solvents, it forms particles which clump together and stick to the walls of the reaction vessel. Moreover, since only the active sites at the surface of the clumps are available for catalysis of the reaction and the sizes of individual clumps vary rather considerably, the overall reaction is rather slow and non-reproducible. No correlation between the amount of enzyme and its reaction could be established.

These problems were overcome by immobilising the enzyme. Since the usually applied methods for immobilisation of the laccase did not work in our system, we had to develop a new method, details of which will be described elsewhere. With this new system, reasonable measurements were possible. Typical patterns of laccase activity could be monitored via the changes of absorbance of the substrates 2,6-dimethoxyphenol and syringaldazine. When the reaction took place in organic solvents, the absorption spectra of the products were similar to those obtained for the same reaction in buffer. Furthermore, the catalytic action of the *T. versicolor* laccase followed Michaelis-Menten-kinetics in most of the organic solvents which were tested (for examples see Table 2).

TABLE 2. Kinetic parameters of oxidation of 2,6 DPM[a] and syringaldezine with Laccase in organic solvents

Substrate	Incubation media	Km mM	Reaction rate[b]
2,6 DMP	Ethyl acetate[c]	0,52	11,6
	Acetonitrile[d]	1,10	12,6
	Acetone[d]	0,65	21,3
	Tetrahydrofurane[d]	1,06	3,4
Syringaldazine	Ethyl acetate[c]	0,15	216
	Acetonitrile[d]	0.08	282
	Acetone[d]	0,08	206
	Tetrahydrofurane[d]	0,31	36

[a] 2,6-dimethoxyphenol
[b] measured as change in absorbance at 468 nm $min^{-1}mg \ E^{-1}$ (2,6 DMP) or $\mu mol \ min^{-1} \ E^{-1}$ (Syringaldazine)
[c] No reaction was observed in the anhydrous solvents. The solvents were presaturated with distilled water at room temperature.
[d] Reaction took place in solvent following additions of 3.5% (v/v) distilled water.

Addition of water always enhanced the efficiency of immobilized laccase. The optimal amounts of water surprisingly were still rather low: rates of syringaldazine oxidation comparable with the ones observed in buffer were, for example, obtained in mixtures 65% aqueous acetonitrile, 50% aqueous acetone and 50% aqueous dioxane (Table 3).

TABLE 3. The rates of Laccase-catalysed oxidation of syringaldazine in different organic solvents

Solvent	Reaction rate μmole min^{-1}mg enzyme^{-1}
Ethyl acetate[a]	216
Toluene[a]	104
Benzene[a]	81
Ether[a]	75
Isoctane[a]	68
n-Hexane[a]	62
Cyclohexane[a]	48
Chloroform[a]	15
Dichlorethane[a]	8
Acetonitrile[b]	116
Acetone[b]	80
Ethanol[b]	28
1.4-Dioxane[b]	16
Tetrahydrofurane[b]	11
Methanol[b]	10
Dimethyl sulphoxide[b]	0
N, N-Dimethylformamide[b]	0

[a] No reaction was observed in the anhydrous solvents. The solvents were presaturated with distilled water at room temperature.
[b] Reaction took place in solvent following additions of 3.5% (v/v) distilled water.

Immobilized laccase has a surprisingly high stability when stored in organic solvents.   Stored in n-hexane at 30°C it was stable for more than eight days with less than a 10% loss of its initial activity, whereas under the same conditions the system lost about 90% of its activity when stored in buffer.   The addition of water to the pure solvents had a detrimental effect on the storage stability of the enzyme.

In summary, data obtained to date indicates that laccase in its immobilized form is a very active enzyme and suitable for any appropriate technical use.

APPLICATION OF LACCASE AS A RADICAL DONOR IN THE PRODUCTION OF ADHESIVES FOR PARTICLE BOARDS

Laccase activity has several attractive features as a candidate for application in technical procedures.   During our investigations, we concentrated on the polymerising activity, attempting to make use of this property for the design of a two-component adhesive with laccase as radical donor and lignin as the recipient molecule;   polymerisation *in situ* being the mechanism for the gluing process.

It is evident that the application of a biological catalyst (the enzyme) in a technical process, in this case the particle board production, demands as "conditiones sine qua non" the following:

i)   the enzyme has to be produced cheaply in large quantities;

ii)   it must be applicable in crude form without prior purification;

iii)   it has to be reasonably stable at room temperature for at least one week;

iv)   since particle boards are pressed at high temperatures, the enzyme must be heat tolerant and have a very high temperature optimum;

v)   the design of the pressing process has to follow the procedure which is applied at present, with as short pressing times as possible;

vi)   the price of the final product has to be competitive with the conventional petrochemical resins which are used today.

Within eight years our research group succeeded in the development of a biotechnological system, which meets all the requirements mentioned above.

The "heart" of the system is a Basidiomycete (*T. versicolor*), grown in a fermenter on waste lignin and a cheap aminophenol as additional enzyme inducing agent. The culture broth obtained after 4 days of cultivation is concentrated by evaporation or ultrafiltration and the resulting concentrated, crude enzyme solution is applied directly as one component in the binding system. Sterile storage of the enzyme concentrate at room temperature is possible for at least one month.

Compared to the chemically catalysed binding processes this biotechnological adhesive technique has certain advantages which are for example:

i)   the total utilization of waste lignin, first as binding component in the adhesive and second as nutrient source for the enzyme production

ii)   because of the high catalytic activity of the enzyme, the binding process can be performed under mild conditions without application of amounts of highly reactive and therefore harmful chemicals

iii)   in contrast to particle boards bonded with synthetic resins the "biobonded" boards do not emit any harmful vapours as do, for example, formaldehyde-bonded boards.

The production process of the biobonded particle boards has been described recently in more detail[1], therefore now will be given only a short summary in favour of an extended presentation of the newest results.

Ball-milled lignin of a certain hydrophobicity is mixed with the crude enzyme solution of *Trametes versicolor* extracellular phenoloxidases produced by cultivation of the fungus on waste lignin, at a ratio of 2:1 (approximately), comprising the main part of the two-component "bio-adhesive". Industrial particles were bonded with 15% bio-adhesive under conventional pressing conditions to give 19 mm particle boards (40 x 50 cm) of the properties described in Table 4).

Table 4 shows that the enzyme has a significant effect on both, the versal tensile strength and the thickness swelling, whereas the lignin type effects mostly the swelling properties of the board: the use of hydrophobic lignin resulted in

TABLE 4. Technological and mechanical properties of "biobonded" particle boards: effect of lignin type, enzyme activity and pressing time

Lignin type	Pressing time (min)	Density (kg/m^3)	Versal tensile strength (N/mm^2)	Thickness after 2 h (%)	Swelling 24 h (%)
1 + enzyme	5	774	0;47	5;4	24.0
2 + enzyme	5	775	0;42	5;4	19.9
2 + enzyme	3	745	0;44	6;7	29.3
1 - enzyme	5	734	0;26	40;1	<60

[a] The lignins were technical by product lignins of different polarity,
   1 = hydrophilic lignin, 2 = hydrophobic lignin.

better water resistance. Pressing time of 3 minutes in combination with enzyme and hydrophobic lignin meets the German standard requirements for V20 particle boards.

CONCLUSION

In general it can be concluded that rather few limitations are met with regard to biological conversion of lignins:

i) there is a variety of enzymes available which can convert the lignin molecule;

ii) different lignin degrading fungi coming from different ecological niches enzymes have very different pattern of these activities. Therefore a suitable organism can be found with "tailor-made" design of enzymes suitable for a particular purpose;

iii) enzymes which act on the lignin molecule introduce new functional groups, especially hydroxyl groups and do not specifically require functional groups for their activity. However, they have a higher activity towards lignins containing high amounts of hydroxyl groups;

iv) lignins extracted from lignocellulosic wastes therefore should be suitable feedstocks for industrial uses of lignins, especially when they are derived from monocotyledons, owing to their higher amount of free phenolic groups in the molecule

v) the main problem which remains is to find methods to extract lignins from wastes in a way that they eventually can compete both with the lignins from the already existing pulping processes and other feedstocks for polymer production which are already on the market.

ACKNOWLEDGEMENTS

The work described in this communication was supported by Grant 86 NR 0063 from the Bundesministerium fur Ernahrung, Landwirtschaft und Forsten, Grant PBE 18938 A from the Bundesministerium fur Forschung und Technologie, and funds from G.A. Pfleiderer, Neumarkt/Opf.

REFERENCES

1. Haars, A.; Trojanowski, J. and Hüttermann, A. (1987).   Lignin bioconversion and its technical application. In: *Bioenvironmental Systems,* Vol. I. Wise, D.L. (ed.) pp. 89-129 Boca Raton: CRC Press.

2. Glasser, W.G. (1981).   Potential role of lignin in tomorrows wood utilization technologies. *Forest Products Journal* **31**: 24-29.

3. Tien, M.; Kirk, T.K. (1983).   Lignin degrading enzyme from the Hymenocete *Phanaerochaete chrysosporium* Burds. *Science* **221**: 661-663.

4. Kirk, T.K. and Farrell, R.L. (1987).   Enzymatic "combustion": the microbial degradation of lignin. *Annual Reviews in Microbiology* **41**: 465-505.

5. Odier, E. (ed.) (1987). *Lignin Enzymic and Microbial Degradation.* INRA Symposia Vol. 40, Paris, INRA Publications.

6. Haars, A. and Hüttermann, A. (1983).   Laccase induction in the white-rot fungus *Heterobasidion annosum* (Fr.) Bref. (*Fomes annosus* Fr. Cooke) *Archives of Microbiology* **134**: 309-313.

7. Ander, P. and Eriksson, K.E. (1976).   The importance of phenol oxidase activity in lignin degradation by the white-rot fungus *Sporotrichum pulverulentum. Archives of Microbiology* **109**: 1-8.

8. Glenn, J.K. and Gold, M.H. (1983).   Decolorization of several polymeric dyes by the lignin degrading Basidiomycete *Phanaerochaete chrysosporium. Applied and Environmental Micorbiology* **45**: 1741-1747.

9. Haars, A.; Majcherczyk, A.; Trojanowski, J. and Hüttermann, A. (1985). Bioconversion of organosolv lignins by different types of fungi. In: *Energy from Biomass.* Palz, W., Coombs, J., Hall, D.O. (eds). pp. 973-977. London and New York: Elsevier.

10. Trojanowski, J.; Milstein, O.; Majcherczyk, A.; Haars, A. and Hüttermann, A. (1987).   Solubilization and Polymerisation of lignin by several wood inhabiting fungi. In: *Lignin Enzymic and Microbial Degradation.* Odier, E. (ed.) INRA Publications.

11. Ishihara, T. (1980).   The role of laccase in lignin biodegradation. In: *Lignin biodegradation: Microbiology, Chemistry, Potential Application.* Kirk, T.K., Higuchi, T. and Chang, H.M. (eds.)   17-31 CRC Press, Inc. Boca Raton, Florida.

12. Hüttermann, A.; Gebauer, M., Volger, C. and Rösger, C. (1977). Polymerisation und Abbau von Ligninsulfonat durch *Fomes annosus*. *Holzforschung* **31**: 83-89.

13. Hüttermann, A.; Herche, C. and Haars, A. (1980). Polymerisation of water-insoluble lignins by *Fomes annosus*. *Holzforschung* **34**: 64-66.

14. Haars, A. and Hüttermann, A. (1980). Macromolecular mechanisms of lignin degradation by *Fomes anosus*. *Naturwissenschaften* **67**: 39-40.

15. Leonowicz, A.; Szklarz, G. and Wojtas-Wasilewska, M. (1985). The effect of fungal laccase on fractionated lignosulfonates (Peritan Na). *Phytochemistry* **24**: 393-396.

16. Ishihara, T. and Miyazaki, M. (1972). Oxidation of milled wood lignin by fungal laccase. *Mokuzai Gakkaishi* **18**: 415-419.

17. Liu, S.Y.; Minard, R.D. and Bollag, J.M. (1981). Oligomerisation of syringic acid, a lignin derivative, by a phenoloxidase. *Soil Science Society of America Journal* **45**: 1100-1105.

18. Lundquist K. and Kristersson, P. (1985). Exhaustive laccase catalysed oxidation of a lignin model compound (vanillyl glycol) produces methanol and polymeric quinoid products. *Biochemical Journal* **229**: 277-279.

19. Kaplan, D.L. (1979). Reactivity of different oxidases with lignins model compounds. *Phytochemistry* **18**: 1917-1919.

20. Martinik, K.; Levashov, A.V.; Klyachko, N.L. and Berezin, J.V. (1977). Water soluble enzymes catalysis in organical solvents (in russian). *Doklady Akademii nauk SSSR* **236**: 920-923.

21. Antonini, E.; Carrea, G. and Cremonesi, P. (1981). Enzyme catalysed reactions in water-organic solvent two-place systems. *Enzyme and Microbial Technology* **3**: 291-296.

22. Butler, L.G. (1979). Enzymes in non-acqueous solvents. *Enzyme and Microbial Technology* **1:** 253-259.

23. Zaks, A. and Klibanov, A.M. (1985). Enzyme-catalyzed processes in organic solvents. *Proceedings of the National Academy of Science of USA* **82**: 3192-3196.

24. Kazandjian, R.Z.; Dorick, J.S. and Klibanov, A.M. (1986). Enzymatic analyses in organic solvents. *Biotechnology and Bioengineering* **28**: 417-421.

25. Dordick, J.S.; Marletta, M.A. and Klibanov, M.A. (1986). Peroxidases depolymerise lignin in organic media but not in water. *Proceedings of the National Academy of Science of USA* **83**: 6225-6257.

26. Lewis, N.G.; Razal, R.A. and Yamamoto, E. (1987). Lignin degradation by peroxidase in organis media: A reassessment. *Proceedings of the National Academy of Science of USA,* **84**: 7924-7927.

COMPARISON OF CHEMICAL AND BIOLOGICAL METHODS FOR PREDICTING FEED INTAKES AND ANIMAL PERFORMANCE

E.R. ØRSKOV and G.W. REID

Rowett Research Institute, Bucksburn, Aberdeen, AB2 9SB, U.K.

SUMMARY

A comparison of biological and chemical measurements to predict intake and growth rate in steers was made with 10 different types of straw including winter barley, spring barley and wheat straw, with and without treatment with ammonia. The biological measurements included determination of metabolisable energy (ME) for sheep in respiration chambers, determination of digestibility in sheep, *in vitro* digestibility and near infrared spectroscopy (NIR) calibrated to *in vitro* and neutral detergent cellulase (NCD) digestibility. In addition each straw was incubated in the rumens of sheep in nylon bags and the rate and extent of degradation was calculated from a mathematical description of degradation with time. Crude fibre, neutral and acid detergent fibre and lignin contents were also determined. Each straw was subsequently given to 8 individually fed steers and these measurements were tested as predictors of intake and growth rate. The correlation of intake and growth rate with chemical measurments was on the whole poor, partly because the chemical measurements showed little change with ammonia treatment. ME and *in vivo* digestibility were poorly correlated with intake and growth rate. *In vitro* digestibility, NIR and, in particular, NCD values were very highly correlated both with intake and growth rate. The best predictor of both intake and growth rate was a multiple correlation based on the degradation rate constant and the potential degradability generated from the mathematical description of degradation with time.

INTRODUCTION

Differences in nutritive value between varieties of straw have been observed by several authors recently.[1-6] However, these authors have mainly used biological methods for identifications. Some of the varietal differences appear to be due to differences in proportions of the botanical fractions of the straw.[4,7,8] Leaf and leaf sheath from straw are generally much more digestible than stems[7,8] except for rice varieties in which the leaf fraction is usually slightly less digestible than stems.[9] Most of the differences between varieties have been due to differences in nutritive value of both stem and leaf and only partly due to differences in leaf:stem ratios.[8] In this paper the ability of different chemical and biological methods to predict the nutritive value of different varities of cereal straw will be discussed. Five varieties were selected on the basis of wide differences in their degradation characteristics and half of the quantity obtained was treated with ammonia to increase the variation so as to enable a more precise evaluation. The results of these experiments have been published in detail elsewhere.[10,11]

RESULTS

The N and ash content of the straw samples were determined, together with their acid (ADF) and neutral detergent fibre (ADF) and acid detergent lignin (ADL) content.[5] Crude fibre was also estimated. The results are given in Table 1. It is noteworthy that the ammonia treatment did not result in consistent differences in chemical fibre analysis though on the whole it tended to decrease lignin and acid detergent fibre content. The N content was, as expected, greatly enhanced by ammonia treatment.

TABLE 1. Chemical composition of different straws offered to sheep either untreated (U) or treated with ammonia (A)

			Composition (g/kg)					
Type	Variety	Treatment	Nitrogen	Ash	NDF	ADF	ADL	CF
Winter	Gerbel	U	7.9	37	875	579	90	458
barley		A	15.8	46	840	555	83	453
	Igri	U	7.6	35	864	554	77	454
		A	14.8	35	841	554	80	450
Spring	Corgi	U	6.7	43	840	512	63	426
barley		A	16.0	41	810	500	61	418
	Golden	U	8.4	48	850	550	73	445
	Promise	A	15.4	48	799	523	69	426
Winter	Norman	U	11.8	61	807	520	80	394
wheat		A	21.1	53	755	488	70	377

The biological measurements including *in vitro* dry matter digestibility and digestibility determined using cellulase enzymes on neutral detergent fibre (NCD) are given in Table 2. The near infrared method (NIR) was also used to estimate digestibility calibrated to *in vitro* digestibility. Measurements obtained with the animals are also given in Table 2. The metabolisable energy (ME) of the feeds were determined with sheep in respiration chambers as described earlier.[10] The *in vivo* digestibility was determined in sheep at the maintenance level of feeding. The voluntary intakes of the straw were measured with the different groups of steers together with growth rate during the period. In addition to the straws the cattle also received daily, 1.5 kg of concentrate to ensure no loss of live weight. A solution of urea was sprayed onto the untreated straw at feeding time to ensure that intake was not limited by N deficiency and the urea added was 20 g/kg straw.

From Table 2 it can be seen that intake of straws differed markedly. Ammonia treatment consistently increased intake, but there were large differences between varieties of straws resulting in live weight gains varying from 106 to 400 g/d with untreated straws and from 332 to 608 g/d with ammonia treated straws.

The degradation characteristics using the nylon bag method and mathematical description by Ørskov and McDonald[12] is given in Table 3. An example of the description and the meaning of each of the factors in the exponential equation $p = a + b (1 - e^{-ct})$ is given in Figure 1. It can been seen that a represents the intercept, b the insoluble but potentially degradable fraction and c the rate constant. It follows that (a + b) represents the maximum digestive potential of the straws or the asymptote of the equation. There was a range in all the values, a varied from 3 to 9, b from 33 to 60 and c from 0.026 to 0.048 and the factors were not intercorrelated so that it was valid to carry out multiple regression analysis to evaluate the ability of the factors individually to predict feed intake and growth rate.

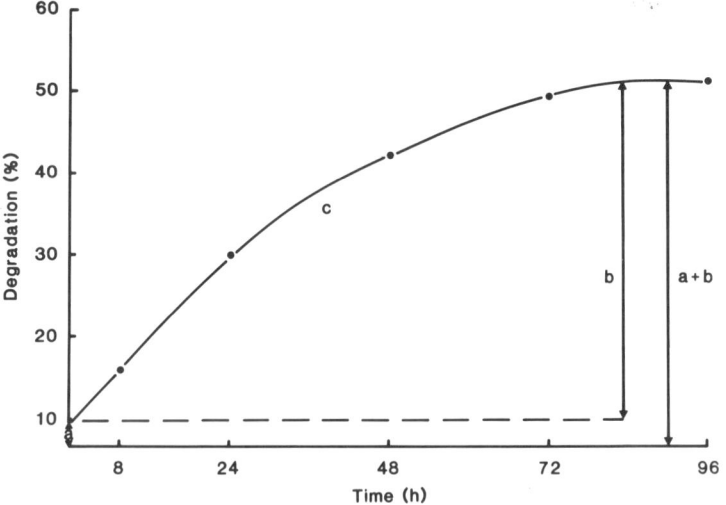

Figure 1. Description of degradation characteristics of straw incubated in the rumen for different intervals of time. The letters are the factors in the exponential equation $p = a + b (1 - e^{-ct})$.

The correlations of the chemical analysis with the animal performance and feed utilisation are given in Table 4. Here it can be seen that on the whole, the correlations were poor. The exception was acid detergent fibre which correlated quite well with intake. This could be expected since the chemical analyses were not consistently affected by ammonia treatment, unlike both straw intake and growth rate measurements. The biological measurements given in Table 4 were consistently more highly correlated with animal performance, particularly, as mentioned before, digestibility determined by NCD, *in vitro* and NIR, the last of which was calibrated with *in vitro* analysis. It is interesting

TABLE 2. The effect of type and variety of straw when fed in untreated (U) or ammonia treated (A) form on growth rate and dry matter (DM) intake by steers, on *in vivo* digestibility and on *in vitro* digestibility. (NCD, neutral detergent cellulase digestibility; NIR, near infrared spectroscopy).

Type	Variety	Treatment	Growth rate (g/d)	Intake of straw (kg DM/d)	*In vivo* digestibility of straw (g/kg)	Metabolizable energy MJ/kg	*In vitro* digestibility (%)	Digestibility derived from: NCD method (%)	NIR (%)
Winter barley	Gerbel	U	106	3.43	409	7.23	27.6	24.7	37.7
	Gerbel	A	359	4.70	487	8.53	37.8	35.3	46.4
	Igri	U	126	3.56	412	7.16	29.5	29.3	41.0
	Igri	A	332	4.82	455	7.90	37.5	34.0	42.9
Spring barley	Corgi	U	400	5.16	484	7.82	39.0	34.4	46.1
	Corgi	A	608	5.86	596	9.50	54.1	47.3	56.9
	Golden Promise	U	198	4.43	452	7.35	36.4	33.5	47.7
	Golden Promise	A	602	4.93	506	8.40	46.5	44.5	51.0
Winter wheat	Norman	U	273	4.57	343	5.98	31.7	31.5	45.3
	Norman	A	516	5.81	484	8.37	44.8	42.9	51.4
s.e.			40	0.17	14				

TABLE 3. The effect of type and variety and ammonia treatment on degradation characteristics at different times of incubation of nylon bags in the rumen and the constants in the exponential equation $p = a + b (1 - e^{-ct})$ (untreated (U) and ammonia treated (A)

Type	Variety	Treatment	Disappearance from nylon bags (g/100 g DM)				a	b	c	Residual s.d.
			7 h	24 h	48 h	72 h				
Winter barley	Gerbel	U	12.9	24.3	33.4	36.0	6.0	32.9	0.0337	0.25
		A	16.9	33.0	46.5	53.8	7.9	54.4	0.0258	1.21
	Igri	U	14.2	28.3	37.4	41.0	5.1	38.2	0.0391	0.97
		A	17.8	33.7	44.8	49.6	7.9	45.2	0.0351	0.63
Spring barley	Corgi	U	17.4	36.9	47.3	50.6	3.4	48.7	0.0483	0.66
		A	22.9	46.6	60.0	64.5	6.4	60.4	0.0457	2.04
	Golden Promise	U	16.6	32.3	44.3	50.1	7.5	48.0	0.0303	1.09
		A	21.4	40.3	52.8	57.9	9.3	52.1	0.0376	0.34
Winter wheat	Norman	U	16.5	30.7	40.8	45.2	7.7	40.9	0.0345	1.76
		A	20.7	39.3	51.9	57.2	9.0	51.9	0.0364	0.81

that both the *in vivo* digestibility with sheep and ME measured at maintenance energy intake were poor predictors of animal performance in spite of the fact that they were the most expensive measurements to make.

TABLE 4. Correlations of chemical and biological measurements with straw intake, growth rates of steers and *in vivo* digestibility measured in sheep

	Dry-matter intake	*In vivo* digestibility	Growth rate
Crude fibre	-0.70	-0.09	-0.57
Neutral-detergent fibre	-0.79	-0.31	-0.77
Acid-detergent fibre	-0.86	-0.45	-0.79
Acid-detergent lignin	-0.75	-0.69	-0.72
Neutral-detergent cellulase digestibility	+0.88	+0.81	+0.95
Near infrared digestibility	+0.86	+0.77	+0.87
In vitro digestibility	+0.89	+0.90	+0.93
Digestibility in sheep	+0.70		+0.77
Metabolisable energy concentration	+0.74		+0.78

The correlations with degradation characteristics derived from nylon bag experiments are given in Table 5. In the first instance the single correlation with the potential (a + b) was tested, then (a + b) + c and finally a, b and c were separated. For the prediction of dry matter and digestible dry matter intake no increase in precision was made by separating a and b. For the prediction of growth rate there was a significant increase in precision when a and b were separated. For all measurments there was a significant improvement in precision by adding the c value (the rate constant).

DISCUSSION

It would seem from this experiment that gross chemical analyses give an inaccurate prediction of animal performance. However there are several promising biological methods. Table 5 would suggest that if we are to develop methods which can also predict voluntary intake the feed value cannot be adequately described by a single value. Several aspects of roughages including the solubility, the potential digestibility and the rate at which the insoluble ingredients are fermented contribute to the feeding value. This is not surprising since these factors impede on rumen fill. The soluble material (a) will occupy little space in the rumen (100 - (a + b)) the totally undegradable material will determine the minimal amounts which will occupy space in the stomach at all times. The rate constant indicates the time during which the degradable portion occupies space in the stomach.

TABLE 5. Prediction of intake of dry matter (DM) and digestible dry matter (DDM) and of growth rate (GR) in cattle from degradation characteristics generated from the equation $p = a + b\,(1\text{-}e^{-ct})$

Factors	Y variable	Formulae	R	Residual s.d.
(a + b)	DM intake kg	0.572 + 0.0766(a + b)	0.83	0.452
(a + b) + c	"	-0.822 + 0.0748(a + b) + 40.7c	0.89	0.375
a + b + c	"	-1.56 + 0.159a + 0.0658b + 56.4c	0.88	0.383
(a + b)	DDM intake kg	1.258 + 0.0642(a + b)	0.86	0.33
(a + b) + c	"	-2.595 + 0.06244(a + b) + 39.0c	0.96	0.195
a + b + c	"	-2.576 + 0.0554a + 0.0640b + 37.7c	0.95	0.204
(a + b)	GR g/d	-0.595 + 0.0175(a + b)	0.84	99
(a + b) + c	"	-0.922 + 0.0170(a + b) + 9.55c	0.91	77
a + b + c	"	-1.267 + 0.0571 + 0.0126b + 17.02c	0.95	54

It was suggested,[10] that if a feed was to be described by a single value it would have to be an index value based on the weighting given to each component such as a, b and c. Since the description of the degradation characteristics appear to be the most reliable it is advisable that methods such as NIR are calibrated against such measurements. It will be interesting to note if characteristics such as solubility, maximum potential and degradation rate can be identified by light reflectance. This could substantially improve the use of these methods to provide rapid ranking of varieties for plant breeding purposes so that quality of straw, whether for industry or animal feeds, can be identified early in plant selection schemes. Recently we have compared NIR with 48 h degradability and obtained R^2 of 0.92 (Murray, Shand and Ørskov, unpublished results). Attempts are presently being made to see whether NIR can predict the rate constants as well.

Of the static biological measurement the NCD method appeared to be particularly interesting however more standardised enzyme mixtures need to be developed. It was also apparent that the 24 h degradation was a more accurate predictor than later incubation times.[10] This may well be expected as such value would provide some measure both of extent and rate. Similar observations were observed several years ago also by Chenost[14] who showed that 24 h degradability was a more accurate predictor of intake than *in vivo* digestibility.

REFERENCES

1. Kernan, J.A.; Coxworth, E.C.; Crowle, W.L. and Spurr, D.T. (1984). The nutritional value of crop residue components from several wheat cultivars grown at different fertilizer levels. *Animal Feed Science and Technology* **11:** 301-311.

2.   Kernan, J.A.; Coxworth, E.C. and Spurr, D.T. (1981). New crop residues and forages for Western Canada: assessment of feeding value *in vitro* and response to ammonia treatment. *Animal Feed Science and Technology* **6:** 257-271.

3.   Kernan, J.A.; Crowle, W.L.; Spurr, D.T. and Coxworth, E.C. (1979). Straw quality of cereal cultivars before and after treatment with anhydrous ammonia. *Canadian Journal of Animal Science* **59:** 511-517.

4.   Bainton, S.J.; Plumb, V.E.; Capper, B.S. and Juliano, B.O. (1987). Botanical composition, chemical analysis and cellulase solubility of rice straw from different varieties. *Animal Production* **44:** 481 (Abstr.).

5.   Pearce, G.R. (1983).   Variability in the composition and *in vitro* digestibility of cereal straws. In *Feed Information and Animal Production* G.E. Robards and R.G. Packham (eds.) Commonwealth Agricultural Bureaux, Farnham Royal, Slough, England.

6.   Tuah, A.K.; Lufadeju, F. and Ørskov, E.R. (1986).   Rumen degradation of straw. 1. Untreated and ammonia-treated barley, oat and wheat straw varieties and triticale straw. *Animal Production* **43:** 261-269.

7.   Ørskov, E.R. and Shand, W.J. (1987).   The effect of type and variety of cereals on nutritive value of straw. In *Straw: A valuaable Raw Material.* Pira Paper Board Division, Leatherhead, Surrey pp 19-30.

8.   Ramanzin, M.; Ørskov, E.R. and Tuah, A.K. (1986).   Rumen degradation of straw. 2. Botanical fractions of straw from two barley cultivars. *Animal Production* **43:** 271-278.

9.   Walli , T.K.; Ørskov, E.R. and Bhargava, P.K. (1988).   Rumen degradability of straw.   3. Botanical fractions of two rice straw varieties and effects of ammonia treatment. *Animal Production* **46:** 347-352.

10. Ørskov, E.R.; Reid, G.W. and Kay, M. (1988).   Prediction of intake by cattle from degradation characteristics of roughages. *Animal Production* **46:** 1, 29-34.

11. Reid, G.W.; Ørskov, E.R. and Kay, M. (1988).   A note on the effect of variety, type of straw and ammonia treatment on digestibility and on growth rate in steers. *Animal Production* **47:** 157-160.

12. van Soest, P.J. and Wine, R.H. (1967).   Use of detergents in the analysis of fibrous feeds. IV. Determination of plant cell wall constituents. *Journal of the Association of Official Analytical Chemists* **50:** 50-55.

13. Ørskov, E.R. and McDonald, I. (1979).   The estimation of protein degradability in the rumen from incubation measurements weighted according to rate of passage. *Journal of Agricultural Science, Cambridge* **92:** 499-503.

14. Chenost, M.; Grenet, E.; Demarquilly, C. and Jarrige, R. (1970). The use of the nylon bag technique for the study of forage digestion in the rumen and for predicting feed value. Proceedings of the 11th International Grassland Congress, Surfers Paradise, pp. 697-701. University of Queensland Press, St. Lucia, Australia.

RECOMMENDATIONS

Much of the available knowledge of plant cell wall composition and molecular architecture derives from studies of primary cell walls. There is a pausity of knowledge of the detailed molecular structure of secondary-thickened plant cell walls; the source of "lignocellulose". This knowledge is needed in such diverse areas as plant breeding, animal husbandry, food processing and biotechnology to optimise the use of renewable biomass and agricultural by-products with a high cell wall content as feed or feedstock.

It is therefore recommended that action should be taken within the Community to:

1. stimulate research which focuses on the molecular characterisation of plant cell walls, their constituent polymers, their ontogeny and their residues after (bio)degradation using modern analytical techniques such as mass spectrometry, infra red spectroscopy and NMR techniques.

2. stimulate research on the molecular topography of the cell wall analysed with modern surface analytical methods in order to understand the effects of topography and topology on the activity of enzyme systems, antibodies and microorganisms.

3. encourage inter-laboratory studies of plant cell walls and their degradation products from which results derived from microscopic, spectroscopic, wet chemical and biological methods can be cross-correlated.

4. encourage the start of a project on the 3-dimensional modelling of polysaccharides and lignin in cell walls by computers using existing and newly acquired data at the monomer and oligomer level.

5. fund and stimulate cooperative research between centres of excellence in Europe and to enable visiting scientists to make use of these facilities.

6. establish and maintain through the medium of COST 84-bis a record of the availability of analytical instruments and techniques of potential value in cell wall research and to use future workshops to make other researchers aware of these possibilities.

LIST OF PARTICIPANTS

AKIN, D.E.
Russell Agricultural Research Center,
P.O. Box 5677,
Athens, Georgia 30613, USA

AMAN, P.
Swedish University of Agricultural
Sciences,
P.O. Box 7024,
S-750 UPPSALA, Sweden

AMARAL COLACO, M.T.,
Laboratório Nacional de Engenharia e
Technologia Industrial (LNETI),
Rua Vale Formoso 1,
1900 LISBOA, Portugal

BENTO, H.,
Laboratório Nacional de Engenharia e
Technologia Industrial (LNETI),
Rua Vale Formoso 1,
1900 LISBOA, Portugal

BOON, J.J.,
FOM Institute for Atomic and
Molecular Physics
Kruislaan 407,
1098 SJ AMSTERDAM,
The Netherlands

CASS, A.,
Imperial College of Science and
Technology,
Prince Consort Road,
LONDON, SW7 2BB, UK

CHESSON, A.,
Rowett Research Institute,
Bucksburn,
ABERDEEN, AB2 9SB, UK

CONCIALINI, V.,
Universitá di Bologna,
Via Selmi 2,
40126 BOLOGNA, Italy

COUGHLAN, M.P.,
Department of Biochemistry,
University College,
GALWAY, Eire

DE WULF, O.,
Faculté de Sciences Agronomique,
Passage des Déportés, 2,
B-5800 GEMBLOUX, Belgium

ENGELS, F.M.,
Wageningen Agricultural
University,
Arboretumlaan 4,
6703 BD WAGENINGEN,
The Netherlands

FORD, C.W.,
Division of Tropical Crops and
Pastures,
CSIRO,
ST. LUCIA, Queensland 4067,
Australia

GALLETTI, G.C.,
University di Bologna,
Via Filippo Re 8,
40126 BOLOGNA, Italy

GIARDINI, A.,
Universitá di Bologna,
Via Filippo Re 8,
40126, BOLOGNA, Italy

GONZALEZ, V.
Consejo Superior de Invest. Cient.,
Institute of Animal Feedingstuffs,
Ciudad Universitaria,
E-28040, MADRID, Spain

GORDON, A.H.,
Rowett Research Institute,
Bucksburn,
ABERDEEN, AB2 9SB, UK

GRENET, E.,
INRA,
Centre de Recherche de
Clermont-Ferrand,
63122 CEYRAT, France

GUADA, A.,
University of Zaragoza,
Departmento Produccion Animal y
Ciencia de los Alimentos,
c/Miguel Serret, 177,
50013, ZARAGOZA, Spain

HIMMELSBACH, D.S.,
Russell Agricultural Research Center,
P.O. Box 5677,
ATHENS, Georgia 30613, USA

HÜTTERMANN, A.,
Forstbotanische Institut,
Universität Göttingen,
GÖTTINGEN, W. Germany

LAPIERRE, C.,
Laboratoire de Chimie Biologique,
INRA,
F78850 THIVERVAL-GRIGNON,
France

LIPPOLIS, M.T.,
Universitá di Bologna,
Via Salmi 2,
40126 BOLOGNA, Italy

LOMAX, J.A.,
Rowett Research Institute,
Bucksburn,
ABERDEEN, AB2 9SB, UK

LÜDEMANN, H.D.,
Institute für Biophysik und
Physikdische Biochemie,
Universität Regensburg,
8400 REGENSBERG,
W. Germany

MONTIES, B.,
Laboratoire de Chimie Biologique,
INRA,
F78850 THIVERVAL-GRIGNON,
France

MORRISON, I.M.,
Hannah Research Institute,
AYR, KA6 5HW, UK

MÜLLER-HARVEY, I.,
Institute for Grassland and Animal
Production,
Hurley,
MAIDENHEAD, SL6 5LR, UK

MURRAY, I.,
North of Scotland College of
Agriculture,
581 King Street,
ABERDEEN AB9 1UD, UK

ØRSKOV, E.R.,
Rowett Research Institute,
Bucksburn,
ABERDEEN, AB2 9SB, UK

PULS, J.,
Bundesforschungsanstalt für Forst
und Holzwirtschaft,
Postfach 80 02 10,
D-2050 HAMBURG, W. Germany

REFFSTRUP, T.,
Bioteknisk Institut,
Holbergsvej 10,
DK-6000 KOLDING, Denmark

REINIGER, P.,
Commission of the European
Communities,
DG XII/E-1,
Rue de la Loi 200,
B-1049 BRUSSELS, Belgium

REXEN, F.,
Carlsberg Research Laboratories
Gamle Carlsbergvej 10,
2500 COPENHAGEN, Denmark

RUSSEL, J.,
Macaulay Land Use Research
Institute,
Craigiebucker,
ABERDEEN, AB9 2QJ, UK

SELF, R.,
Rowett Research Institute,
Bucksburn,
ABERDEEN, AB2 9SB, UK

STEWART, C.S.,
Rowett Researtch Institute,
Bucksburn,
ABERDEEN, AB2 9SB, UK

VAN DER MEER, J.M.
Institute for Livestock Feeding and
Nutrition Research (IVVO),
P.O. Box 160,
8200 AD LELYSTAD,
The Netherlands

WILLEMSE, M.T.M.,
Wageningen Agricultural University,
Aboretumlann 4,
6703 BD WAGENINGEN,
The Netherlands

ZADRAZIL, F.,
Bundesforschungsanstalt für
Landwirtschaft,
Institute für Bodenbiologie,
Bundesallee 50,
D-3300 BRAUNSCHWEIG,
W. Germany

ZIMMERMANN, W.,
Eidgenössische Technische
Hochschule,
ETH-Hönggerberg
8093 ZURICH, Switzerland.